Explaining an Eclipse

Explaining an Eclipse

Aristotle's *Posterior Analytics* 2.1–10

Owen Goldin

Ann Arbor

THE UNIVERSITY OF MICHIGAN PRESS

Published in the United States of America by
The University of Michigan Press
Manufactured in the United States of America
⊗ Printed on acid-free paper

1999 1998 1997 1996 4 3 2 1

A CIP catalog record for this book is available from the British Library.

Library of Congress Cataloging-in-Publication Data

Goldin, Owen, 1957–
Explaining an eclipse: Aristotle's Posterior analytics 2.1–10 / Owen Goldin.
p. cm.
Includes bibliographical references and index.
ISBN 0-472-10596-5 (alk. paper)
1. Aristotle. Posterior analytics. 2. Definition (Logic). 3. Science—Philosophy. 4. Knowledge, Theory of. 5. Ontology. I. Title.
B441.G65 1996
160—dc20 95-50076
CIP

Dedicated to my wife

Preface

The first ten chapters of the second book of Aristotle's *Posterior Analytics* make up a single discussion whose apparent purpose is to clarify the relations that hold among demonstration, definition, existence, and essence, all important features of scientific explanation as Aristotle conceives of it. The whole discussion is explicitly concerned with a single problem: whether a demonstration of a definition is possible. *Posterior Analytics* 2.3–7 present *aporiai* (conceptual difficulties) to the effect that a demonstration of a definition is impossible. *Posterior Analytics* 2.8–9 purport to present a way in which it is possible, and 2.10 is something of a summary. But because Aristotle's prose is never so opaque as it is in 2.8–9, Aristotle scholars have reached no consensus on what this way is and what significance it has. Accordingly, these chapters have been of peripheral interest among even those specializing in Aristotle's philosophy of science. They have often been treated not as a careful philosophical discussion aimed at resolving a single problem, but as a collection of disparate observations of varying interest. It is small wonder that a commentator as astute as Barnes can say of the extremely important 2.8, "The argument is tortuous, and I can find no way of extracting a single train of reasoning from it."[1]

What is needed is an interpretation of these chapters that is both well supported by the text, and that allows us to see Aristotle addressing an important problem in a unified and coherent manner. Within this study, the first book-length treatment of these chapters in recent times, I have tried to present and defend such an interpretation. The general line of interpretation I adopt is that of W.D. Ross and, more recently, B. Landor.[2] But their discussions of 2.1–10 are brief, leaving crucial questions unanswered and leaving unexplored the issue of the chapters'

1. J. Barnes, *Aristotle's "Posterior Analytics," Translated with Notes* (Oxford: Clarendon Press, 1975), 208.

2. W.D. Ross, *Aristotle's "Prior and Posterior Analytics"* (Oxford: Clarendon Press, 1949), 75–78, 609–36; B. Landor, "Aristotle on Demonstrating Essence." *Apeiron* 19 (1985): 116–32.

philosophical significance. I have tried to supply what is lacking in their exposition.

I argue that within *Posterior Analytics* 2.8–9 Aristotle is showing how the existence or occurrence of some entity, attribute, or event that is not among the theoretical primitives of the sciences can be explained on the basis of such primitives. Aristotle explains how such an explanation can be accommodated within his syllogistic and shows how it is that such an explanation allows one to express the essence of this entity, attribute, or event in a scientifically adequate definition. This interpretation has a number of interesting results. Three are as follows. First, this interpretation shows that Aristotle is not mistaken in taking demonstrations (cast in the mold of his syllogistic) to be able to do more than render explicit genus-species relations. Second, it calls into question the extent to which Aristotle is committed to the autonomy of the particular sciences. Third, it shows that Aristotle has a rather sophisticated account of the sort of explanation that is today called reductive.

I owe an immense debt to the two people who helped me the most in this book's early stages,[3] Alexander P.D. Mourelatos and Sarah Broadie. In their comments and questions, as well as through the example of their own work, they have set for me the highest standards of scholarship and philosophical thinking. Thanks are also due to Daniel Flage, Cynthia Freeland, Henry Mendell, and all of the others who have commented on this project along the way and with whom I have discussed the ideas presented within it. A special debt of gratitude is owed to Thomas Upton, Michael Ferejohn, and to an anonymous reviewer. They have read an earlier version of this book with great care. It is a much better book for their philosophically sensitive and incisive comments.

My teachers, friends, and fellow scholars have kept me from making some mistakes; I am to blame for any in which I have persisted.

I regret that the Jonathan Barnes' revision of his seminal Clarendon edition of the *Posterior Analytics*[4] was available to me only at the latest stages of the preparation of this manuscript. References are to the first edition, though it has in part been superseded.

Some of the work on this book was supported by a Summer Faculty

3. O. Goldin, "The Role of the Demonstration That Explains an Essence in Aristotle's Theory of Explanation" (Ph.D. diss., University of Texas at Austin, 1987).

4. J. Barnes, *Aristotle's "Posterior Analytics," Translated with a Commentary*, 2d ed. (Oxford: Clarendon Press, 1994).

Fellowship from Marquette University. Three research assistants provided by Marquette University, Calvin Pafford Jr., Eric Manchester, Mark Munns, and Warren von Eischenbach have helped with proofreading, indices, and editing.

Many thanks are also due to Ellen Bauerle and Nancy Vlahakis, of the University of Michigan Press, for their support, guidance, and patience.

My wife, Patricia Kilroe, helped enormously with editing. There are many more important ways in which the book is a product of our life together. With love, I dedicate it to her.

Contents

Abbreviations xiii

1. A Problem in Aristotle's Theory of Scientific Demonstration 1

2. Aristotle on the Objects of Explanation 15

3. Epistemic Substances and Assumptions of Existence 41

4. The Aporiai 79

5. Inquiry and Explanation: *Posterior Analytics* 2.8 101

6. On Demonstrating καθ' αὑτὰ συμβεβηκότα Propositions 137

Works Cited 157

Index of Names 163

Index Locorum 165

Abbreviations

APo.	*Posterior Analytics*
APr.	*Prior Analytics*
Cat.	*Categories*
DC	*On the Heavens*
De An.	*On the Soul*
DI	*On Interpretation*
EE	*Eudemian Ethics*
EN	*Nicomachean Ethics*
Metaph.	*Metaphysics*
Mete.	*Meteorology*
PA	*Parts of Animals*
Phys.	*Physics*
Pol.	*Politics*
Poet.	*Poetics*
Rep.	*Republic*
SE	*Sophistical Refutations*
Top.	*Topics*

1

A Problem in Aristotle's Theory of Scientific Demonstration

A. How Are Demonstrations of καθ' αὑτὰ συμβεβηκότα Predications Possible?

Within the *Posterior Analytics* Aristotle presents his theory of demonstration, the kind of explanation of a fact that allows one to have ἐπιστήμη (scientific understanding)[1] of that fact. Much of this work is devoted to clarifying the syllogistic form of demonstrations.[2] But, as

1. M.F. Burnyeat convincingly argues that *understanding* best captures the sense of ἐπιστήμη, as employed by Aristotle ("Aristotle on Understanding Knowledge," in *Aristotle on Science: Proceedings of the Eighth Symposium Aristotelicum*, ed. E. Berti [Padua: Editrice Antenore, 1981]).

2. As has been often noted, Aristotle's own scientific explanations are rarely cast in syllogistic form; this has led readers of Aristotle to question the extent to which he is committed to the view that all demonstrations can be cast in the form of syllogisms. Barnes argued that the syllogistic form of demonstrations is taken by Aristotle to be the optimal expression of explanations in the context of instructing students ("Aristotle's Theory of Demonstration," *Phronesis* 14 [1969]). Several years later Barnes suggested another possibility: the core of Aristotle's theory of demonstration is independent of the syllogistic, which postdates both this core and Aristotle's scientific writings ("Proof and the Syllogism," in *Aristotle on Science: Proceedings of the Eighth Symposium Aristotelicum*, ed. E. Berti [Padua: Editrice Antenore, 1981]). A third proposal, which is a development of Barnes' thought, has gained currency among many scholars. This is to the effect that Aristotle takes the syllogistic expression of demonstrations to make clear the logical structure of all demonstrations, though these latter need not be explicitly expressed in syllogistic form in the context of scientific discourse. On this, see Burnyeat, "Aristotle on Understanding Knowledge," 97–120; M.F. Burnyeat and J. Barnes, "Socrates and the Jury: Paradoxes in Plato's Distinction between Knowledge and True Belief," *Aristotelian Society*, supp. 54 (1980): 188, 204–5; A. Gotthelf, "First Principles in Aristotle's *Parts of Animals*," in *Philosophical Issues in Aristotle's Biology*, ed. A. Gotthelf and J. Lennox (Cambridge: Cambridge University Press, 1987), 167–70, 194–97; J.G. Lennox, "Divide and Explain: The *Posterior Analytics* in Practice," in Gotthelf and Lennox, *Philosophical Issues*, 118–19; and M. Ferejohn, *The Origins of Aristotelian Science* (New Haven: Yale University Press, 1991), 141 n. 4. I subscribe to this third view. If the second, developmental view is correct, then the problem I am about to present in this chapter does not arise, at least in the core of Aristotle's theory, which predates the theory of the syllogism. *APo.* 2.8–10, which in my view is concerned with a problem Aristotle recognizes with giving demonstrations syllogistic form,

has often been remarked, on account of its limited power, Aristotelian syllogistic cannot fill its intended role as the formal framework for the theory of explanation developed in the *Posterior Analytics*. The mathematical sciences are a prime example of this; although Aristotle often employs mathematical examples within the *Posterior Analytics*, mathematical proofs almost always require multiplaced predicates and patterns of inference that cannot be sanctioned by Aristotelian syllogistic.[3] It is not often recognized, however, that, as it is usually interpreted, the syllogistic structure of Aristotle's theory of explanation binds scientific explanation so radically that it excludes every nontautologous proposition from scientific explanation.

This is why. Aristotle argues that if scientific explanations are to be ultimate, as he takes them to be, they must be grounded in certain predications that do not themselves demand explanation. These are referred to as *atomic*, or *unmediated*, predications and are culled from the first principles of each of the sciences (1.2.72a5–8, 3.72b19–24, 15.79a34–36; 2.19.99b20–22). Aristotle's classification of the kinds of first principles (definitions, hypotheses, and axioms [cf. 1.2.72a14–24, 10.76a31–77a4]) is notoriously unclear and has given rise to considerable debate. Within the next chapter I present my view of the distinction between definitions and hypotheses, but at this point it is enough to note that the only kind of principle discussed by Aristotle that can furnish unmediated predications (in which some single term is predicated of x) will be a *definition* of x.[4] Since demonstrations are first figure universal syllogisms,

---

would on this view not be interpreted as concerned with the syllogistic at all. There are two responses to this. First, no more cogent interpretation of 2.8–10 has yet been presented. Second, the series of aporiai (conceptual difficulties) that precede 2.8–10 lean heavily on Aristotle's theory of the syllogism. This indicates that the theory of the syllogism plays an important role in the problem that Aristotle sets out to solve.

3. The theory of demonstration of *APo.* is presented as applicable to all the sciences. But at *APr.* 1.35.48a29–39 Aristotle shows that he is aware of demonstrative λόγοι that are not syllogisms, for he indicates that the proposition that triangles have the sum of their interior angles equal to two right angles is demonstrable, though there is no τόδε τι (some this) that functions as a syllogistic middle term. On this, see R. Smith, *Aristotle: "Prior Analytics"* (Indianapolis: Hackett, 1989), 164–65. For a discussion of why Aristotelian syllogistic is unsuitable for formalizing geometrical deductions, see I. Mueller, "Greek Mathematics and Greek Logic," in *Ancient Logic and Its Modern Interpretations*, ed. J. Corcoran (Dordrecht, Neth.: D. Reidel, 1974), 37–57.

4. This is because the requisite universal predications can be derived neither from assertions of existence [Aristotelian hypotheses as traditionally interpreted (see J. Barnes, *Aristotle's "Posterior Analytics," Translated with Notes* [Oxford: Clarendon Press, 1975], 104)]

the explanation of any proposition of the form S is P must in part rest on a member of that set of predications that makes up the essence of S. This is at least part of what Aristotle is saying when at 2.2.90a14–15 the "what is it" and the "why is it" are identified; we understand why S is P by seeing how that logically follows from what x is.[5]

In the *Organon* as well as throughout the rest of the Aristotelian corpus, definitions are taken to have the form of the genus conjoined with the differentia.[6] This means that all of the unmediated atomic predications that are the basis of Aristotelian demonstrations will have one of the following two forms: "All K is D" or "All K is G" (where K is a kind, G is the genus of K, and D is the differentia of K). G itself may be definable as the conjunction of a superordinate genus G′ and differentia D′, and so on until a highest genus is reached. (Aristotle does not, within the *Posterior Analytics*, speak of definitions of differentiae; it is the genera, or kinds, treated by a science, whose essences are assumed as first principles [1.10.76b3-6].[7] But, if the differentia D is defined, it would be

---

nor from statements that identify the kinds under which particulars fall. Possible exceptions to this, unmediated predications that are not expressed in the definitions of their subjects, are what Aristotle calls "common principles" or "axioms." These are logical truths and those mathematical truths that are common to more than one kind of magnitude (such as "when equals are subtracted from equals the results are equal" [see 1.10.76a41–b2, 76b14–15; 11.77a26–32]). Thus, McKirahan takes common principles or axioms to be kinds of basic premises (*Principles and Proofs: Aristotle's Theory of Demonstrative Science* [Princeton: Princeton University Press, 1992], 75).

5. On this see chap. 2, sec. A.

6. See *Top.* 7.3.153a15–18; *APo.* 2.13; and *Metaph.* 7.12.1037b28–1038a4. In *PA* 1.2 and 3 Aristotle's theory of definition is more subtle than elsewhere; he rejects the requirement that corresponding to each kind there be a *single* simple differentia that sets that kind off from others belonging to the next most inclusive genus and recognizes that often it is necessary for the definition to contain a conjunction of various attributes, each an independent differentia. (On this, see D. Balme, *Aristotle's "De Partibus Animalium" I and "De Generatione Animalium" I* [Oxford: Clarendon Press, 1972], 105; P. Pellegrin, *Aristotle's Classification of Animals*, trans. A. Preus [Berkeley: University of California Press, 1986], 13–49). McKirahan (*Principles and Proofs*, 111–21) argues that this innovation provides the means for solving the problem I will outline, for a definition with multiple differentiae could implicitly express everything about the subject that can be scientifically explained. I agree with McKirahan and with the Aristotle of *PA* that any definition that could serve as the basis for all possible demonstrations about its subject would almost always require multiple independent differentiae. But, for reasons to be made clear, I do not believe that this innovation alone renders Aristotle's account unproblematical.

7. However, *Metaph.* 8.6.1045a33–b7 can be interpreted (in my opinion, rightly) to the effect that a differentia cannot be analyzed as a genus and a differentia. See L.M. De Rijk, *The Place of the Categories of Being in Aristotle's Philosophy* (Assen: Van Gorcum, 1952), 61–62.

through identifying a new genus and differentia; both of these could conceivably be defined, generating more genera and differentiae, and so on.) That there would be an end to such a regress is clear from *Posterior Analytics* 1.22.82b37–83a1 and 83b1–3, in which Aristotle argues that the fact that an essence can be known, together with the premise that an indefinite sequence is unknowable, entails the finitude of the number of predicates that make up that essence.[8] All of these definientia can be considered as implicitly present within the definition of K. Let us call all such predicates (including D and G themselves) *implicitly definitional* of K. For every kind K there will be a set S ({G, D, G′, D′, . . .}) of all of those predicates that are implicitly definitional of K. For example, in the case of "human being" this set would have as members "animal" and "rational" and also the genus and differentia of "animal" as well as the genus and differentia of "rational" (if differentiae are defined), the genera and differentiae of *these* genera (and, again, of the differentiae, if they can are defined), and so on.

If demonstrations are to have the form of the first figure universal syllogism (1.14, 24), and if all of the ultimate premises of demonstration are definitional, it is easily shown that the only predicates demonstrable of K are members of S. In other words, according to Aristotle's theory, as interpreted in this way, a demonstration is limited to rendering genus-species relations explicit.

But Aristotle takes his theory of explanation to be much more powerful than this. Even if we leave aside the scientific explanations Aristotle puts forward outside of the *Posterior Analytics* and the mathematical examples employed within it, examples abound of projected demonstrations of predications that are not already implicit within the definition of the subject. For example, "the planets do not twinkle" (1.13.78a39–b4), "the Moon is sometimes eclipsed" (2.2.90a14–23, 8.93b2–7, 16.98b16–24), "the clouds sometimes thunder" (2.8.93b7–14, 10.93b38–94a9), "deciduous plants shed their leaves" (2.16.98b5–16, 17.99a23–29), "quadrupeds live a long time" (2.17.99b5–6), "birds live a long time" (2.17.99b5–7), and "the high and low note harmonize" (2.2.90a18–23) are all taken by Aristotle to be demonstrable.

Further, Aristotle explicitly recognizes a class of attributes, those that are καθ' αὑτὰ συμβεβηκότα or "per se incidental" in relation to a

8. On this, see Barnes' exposition of this passage (which implicitly assumes that differentiae are not to be defined) (Barnes, *Aristotle's "Posterior Analytics,"* 166–67, 170).

kind, which are demonstrable of that kind but not present in the definition.[9] These are καθ' αὑτά insofar as they inhere in the subject on account of the subject (cf. *Metaph.* 5.18.1022a14–22)[10] but are incidental insofar as they are not present in the definition of the subject (cf. *APo.* 1.4.73b4–5, 22.83b17–20; *Metaph.* 5.30.1025a30–32).[11] These attributes are demonstrated to inhere in the subjects studied by the sciences (*APo.* 1.7.75a42–b2; cf. 1.10.76b12–15). One way out of the difficulty would be to argue that the καθ' αὑτὰ συμβεβηκότα of K are those members of S that are neither the genus nor the differentia of K (the two predicates explicitly present in K's definition). For example, on this understanding the genus of the genus of K would be a καθ' αὑτὸ συμβεβηκός predicate of K.[12] But, although Aristotle presents a number of examples of demonstrable, καθ' αὑτὰ συμβεβηκότα predications, none are implicitly definitional. The text offers no support for taking those predicates that are καθ' αὑτὰ συμβεβηκότα of K to be members of S.

Either Aristotle is seriously mistaken concerning the explanatory power of his theory of demonstration or this theory is more complicated and supple than the usual interpretation would have it.

B. The Solutions of Ferejohn and McKirahan

To the best of my knowledge, M. Ferejohn and R. McKirahan have been the only scholars to recognize these difficulties. They recognize that, if first principles either predicate a genus of a subject[13] or a differentia of a subject,[14] the predications that Aristotle would take to be subject to

9. For the sake of simplicity, I shall employ the phrase καθ' αὑτὸ συμβεβηκός in regard to both attributes and predications. An attribute is καθ' αὑτὸ συμβεβηκός in respect to some subject if and only if the predication of that attribute of that subject is καθ' αὑτὸ συμβεβηκός.

10. On this, cf. chap. 7 n. 9.

11. See W. Kullmann, "Die Funktion der mathematischen Beispiele in Aristotles' *Analytica Posteriora*," in *Aristotle on Science: Proceedings of the Eighth Symposium Aristotelicum*, ed. E. Berti (Padua: Editrice Antenore, 1981), 266–67.

12. That the καθ' αὑτὰ συμβεβηκότα include such members of K is suggested by M. Ferejohn, "Definition and the Two Stages of Aristotelian Demonstration," *Review of Metaphysics* 36 (1982): 395.

13. Ferejohn argues that such a predication is that called καθ' αὑτό at 1.4.73a35–38; he calls it a type 1 per se predication. Ferejohn, *Origins*, 89–91.

14. Ferejohn (ibid., 96–108) argues that this kind of predication is that called καθ' αὑτό at 1.4.73a38–b4; he calls this a type 2 per se predication. I am not convinced by Ferejohn's account of the kinds of καθ' αὑτό predication discussed in 1.4, but here there is no need to arrive at a definitive interpretation of this difficult chapter.

scientific explanation would be "severely restricted to what we might now call analytic truths." Ferejohn writes:

> From the standpoint of modern philosophers of science, a theory of scientific explanation that took such a narrow view of the scientific domain would be utterly intolerable, since it would exclude precisely the sort of statement with which scientists are most concerned: those logically contingent, but highly probable, generalizations thought to express natural causal connections. Indeed, the only kind of premise countenanced by such a theory, those whose truth follows from the definitions of the terms they contain, are the concerns of such exact disciplines as mathematics and logic,[15] which are generally classified as sciences only in a very special sense of the term.[16]

Both Ferejohn and McKirahan suggest solutions to this problem, by which Aristotle's theory of explanation can indeed accommodate explanations of predications that are not implicitly definitional. I here argue that these accounts are inadequate and that for this reason the theory of explanation presented in *Posterior Analytics* book 1 is incomplete in a crucial respect. In the following chapters I argue that this gap is filled by *Posterior Analytics* 2.1–10.

Ferejohn's proposal rests on his interpretation of the kind of predication called καθ' αὑτό at *Posterior Analytics* 1.4.73b10–16, which he calls type 4 per se predication. On Ferejohn's account these include two related kinds of predications found in the sciences. The first is ἐπὶ τὸ πολύ (for the most part) predications, which express "causal generalizations." These will be those causal connections such as "a man grows chin whiskers as he ages" (*APo.* 2.12.96a10), which, though admitting some exceptions, can be the object of scientific understanding.[17] Fere-

---

15. Of course, as already noted, even if mathematical first principles are of the form of genus and differentia, the proofs for which they form the basis can rarely be cast in the mold of Aristotelian syllogistic.

16. Ferejohn, *Origins*, 177. The problem is also isolated and discussed in Ferejohn, "Definition," 390 n. 37. The solution to the problem that Ferejohn advances in his paper (in an explicitly tentative manner) differs in important respects from that presented in his book, which presumably supersedes his earlier work. I here discuss only the solution that Ferejohn presents in his book. For another exposition of the problem, see McKirahan, *Principles and Proofs*, 111–15.

17. Ferejohn, *Origins*, 119–23.

john takes as an example of this sort of predication that which Aristotle explicitly presents at *Posterior Analytics* 1.4.73b14–16 ("death" is predicated of "animal having its throat cut" καθ' αὑτό, for the animal dies *because* its throat is cut).[18] The other kind of type 4 per se predication is the predication of a καθ' αὑτὸ ἴδιον (*per se proprium*) of a subject. This is a predication such as "triangles have interior angles equal to two right angles," in which the predicate belongs only to the subject, and in which there is a causal connection between the subject and predicate.[19] If such predications were to be admitted as among the first principles of the sciences, it is clear that there could be demonstrations whose conclusions are not implicitly definitional.

I will not here discuss whether this is indeed an accurate account of the kind of καθ' αὑτό predication discussed at 1.4.73b10–16. It is indisputable that Aristotle does take ἐπὶ τὸ πολύ and καθ' αὑτὰ ἴδια predications to be among the objects of scientific understanding. But what status will they have within the formal structure of the sciences? It is clear that at least some of them, such as "triangles have interior angles equal to two right angles," will be among the conclusions of demonstrations.[20] But, for Ferejohn's solution to work, there must be others that have the status as ultimate, atomic predications. But what is at issue is how Aristotle's theory can accommodate this. Is there evidence that expressions of these causal connections will be among the basic premises, or first principles, of the sciences?

Plainly, these sorts of predications can and do serve as demonstrative premises. For example, any geometrical καθ' αὑτὰ ἴδια predication that is already proved can be employed as a premise in a further proof. Without this there could be no substantial body of geometrical knowledge, such as pre-Euclidean geometry, which Aristotle frequently takes as a model of a demonstrative science. The point also holds outside of geometry.

Recent research by Lennox on Aristotle's biological explanations shows how they conform to the basic structure of demonstration as envisaged in the *Posterior Analytics*.[21] Lennox first points to examples of what he calls A-type explanations, in which a predication is explained

18. Ibid., 119.
19. Ibid., 123–25. Cf. also Ferejohn, "Definition," 391 n. 37.
20. This is recognized in Ferejohn, *Origins*, 124.
21. Lennox, "Divide and Explain," 90–97.

by its being shown that the major term (the predicate) belongs to some class (the middle term), which is more encompassing than the subject (the minor term). A geometrical example is:

(i) All isosceles triangles are triangles,
(ii) all triangles have interior angles equal to two right angles,
so, (iii) all isosceles triangles have interior angles equal to two right angles.

Lennox distinguishes such a demonstration from one he calls a B-type demonstration, in which a predication is explained by its being shown that the major term belongs to some class that is coextensive with the subject. An example is that alluded to at 2.16.98b5–16:

(i) Flat-leaved plants are plants such that the sap solidifies at the juncture of the seed,
(ii) any plant such that the sap solidifies at the juncture of the seed sheds its leaves,
so, (iii) flat-leaved plants shed their leaves. (Cf. *APo.* 2.17.99a25–29)

In such a case premise ii, which is the major premise (the predication of the major term of the middle term) is a καθ' αὑτὸ ἴδιον predication. The conclusion, iii, is demonstrated only by virtue of the role of ii as a premise. Ferejohn appeals to Lennox's discussion of B-type explanations to show how type 4 per se predications can function as demonstrative premises.[22] With such premises demonstrations of καθ' αὑτὰ συμβεβηκότα predications become possible.

But it is not enough to show that the demonstrations of καθ' αὑτὰ συμβεβηκότα predications can have ἐπὶ τὸ πολύ and καθ' αὑτὰ ἴδια predications as premises. It must be shown exactly how this is the case. There are two possibilities. One is that such a premise is itself demonstrable. In this case the problem is merely deferred, for we must wonder how it is the case that these premises are demonstrable. Again, it does not seem to be the case that all of the necessary premises will be implicit within a genus/differentia definition of the subject. The second possibility is that these premises are themselves first principles. Again,

22. Ferejohn, *Origins*, 125–28, 160–61 n. 22.

this does not seem compatible with Aristotle's apparent belief that all atomic premises are implicit within genus-differentia definitions.

To clarify matters, let us return to the example of a type-B demonstration, focusing on the role of premise ii. (Ferejohn might also classify premise i as a type 4 per se predication, but it could conceivably be implicit in a genus-differentia definition of "flat-leaved plant.") Is premise ii demonstrable? If so, how? Is it implicitly definitional? In that case the definition of "plant such that the sap solidifies at the juncture of the seed" would need to have the status of being a biological definition, and the predicate of "shedding leaves" would need to be implicit within it. But it seems an unlikely and ad hoc move to make "shedding" part of the essence of "plant such that the sap solidifies at the juncture of the seed." The alternative would be to say that it is *demonstrable* that plants whose sap solidifies at the juncture of the seed shed their leaves. But then we return to the question of how such a demonstration is possible in the first place.

The upshot is that type 4 per se predications, as conceived by Ferejohn, can indeed serve as demonstrative premises but not as ultimate demonstrative premises or first principles. We should suspect as much from the example of a type 4 per se predication that Ferejohn finds in the text: "death" is predicated of "animal having its throat cut." How could this possibly be an example of a first principle? If the science of biology is unable to explain why an animal whose throat is cut dies (i.e., because it is no longer able to breathe), there is little it could explain.

It would seem that, if all demonstrative premises are indeed to be derived from definitions, the reader of the *Posterior Analytics* must bite the bullet and take all premises of demonstration to be implicitly definitional, even though this runs counter to the notion of a definition as a conjunction of a genus and a single differentia. This is the interpretive path taken by R. McKirahan.

According to McKirahan, there are two conceptions of definition operative within the *Posterior Analytics*.[23] One is that of a single genus and differentia, prominent in 1.22 and 2.13. The other is never explicitly discussed within the *Posterior Analytics*, or anywhere else, for that matter, but is rendered necessary by Aristotle's conception of demonstra-

23. I disregard a third kind of definition, which McKirahan discusses (*Principles and Proofs*, 113–15), that which he calls "type B" definitions. This is the sort of definition described in 2.13.96a24–b14. As McKirahan points out, this conception of definition is anomalous; hence, it is not germane to the present discussion.

tions as based on ultimate premises derived from definitions. On this understanding the definition of a subject is a conjunction of all of the atomic predications of that subject.

> Since the only kinds of principles are definitions, axioms, and existence claims, the definition of a thing must contain all the essential facts about it, from which (together with the other principles of the science) the derivative facts about it are proved. The definition of a subject or attribute, *x*, will in general be a compound statement of the form "x is A_1 & x is A_2 & . . . & x is A_n," where each of the "x is A_i" statements represents an immediate per se relation. I shall call this a "fat definition."[24]

McKirahan accepts the implications of his position, asserting in respect to the biological demonstration sketched here that being deciduous (shedding) "is part of the fat definition of plant-in-which-the-sap-congeals-in-the-connection-of-the-leaf-stalk."[25]

This solution of the problem cuts a Gordian knot. Scientific explanation is indeed reduced to explicating analytic truth; this is made possible by packing an enormous amount of information about a subject into its scientific definition. I believe that Ferejohn is correct; such a conception of scientific explanation is not nearly as philosophically interesting as Aristotle's theory promises to be.

In wondering how the demonstration of καθ' αὑτὰ συμβεβηκότα predications can be accommodated by Aristotle's syllogistic, Ferejohn and McKirahan have called attention to an important problem faced by Aristotle's theory of demonstration as it is usually interpreted. But they have not been able to solve it. In the following section I sketch the solution that I will present at length in this book.

C. An Outline of a Solution

The solution to this problem is presented in the first ten chapters of the second book of the *Posterior Analytics*. Readers of the *Posterior Analytics*

---

24. Ibid., 111. Shortly after this passage McKirahan identifies a fat definition with the definition advocated in *PA* 1.2–3, a conjunction of a genus and multiple independent differentiae. (On this, see n. 6.)

25. Ibid., 169.

have seldom recognized this because Aristotle himself presents these chapters as unified by a concern with what appears to be a very different problem: whether there can be a demonstration of a definition. After a preliminary account of the kinds of questions that can be the concern of a science (2.1–2), Aristotle proceeds to marshal out arguments to show that there can, in fact, be no demonstration of a definition (2.3–7). Aristotle then presents an account of how, in either a special sense of *demonstration*, a special sense of *definition*, or in a special sense of both terms, there can in fact be such a demonstration (2.8–10). Scholars have found Aristotle's arguments in these three chapters exceedingly obscure, and there is no consensus on what Aristotle thinks he has shown and on whether his account is successful.

This is to a large extent because the problem that Aristotle sets for himself is, on the face of it, not problematical at all. To the question of whether there can be a demonstration of a definition, the careful reader of the first book of the *Posterior Analytics* would reply with an unequivocal no. For the upshot of that book is that demonstration is a kind of explanation that shows why certain universal and necessary facts follow as a matter of logical necessity from unmediated premises, which themselves neither demand nor accept explanation. Definitions, being among the first principles, are conjunctions of such unmediated premises. Accordingly, a demonstration of a definition would be an explanation of the unexplainable.

Since in *Posterior Analytics* 2.8–9 Aristotle does grant that there is a sense in which there is a demonstration of a definition, he must be granting a new sense to the term *demonstration*, or to *definition*, or to both. But he never makes explicit which terms are given a new sense and what this sense is. In the chapters that immediately precede *Posterior Analytics* 2.8–9, however, Aristotle does provide the needed direction.

A main theme of 2.1–2 is the logical relationships that hold among essence (τὸ τί ἐστί), existence (τὸ εἶναι ἁπλῶς), and being the subject of a predication (τὸ εἶναι ἐπὶ μέρους). On my interpretation of *Posterior Analytics* 2.1–2, Aristotle's analysis of these relationships is the theoretical framework for his solution to the problem of the form of the demonstration of καθ' αὑτὰ συμβεβηκότα predications.

I present my account of Aristotle's analysis in chapter 2. I argue that for the two very different kinds of beings with which scientific explanation deals, Aristotle is committed to very different analyses of these relationships. The subjects of the sciences make up one of these kinds. These include, but are not limited to, those kinds that are substantial from a metaphysical point of view. They have in common the feature of not being considered by the sciences studying them to have their existence dependent on their inherence in any more fundamental kind. I call them "epistemic substances." The other kind is made up of those features of the world that *are* considered by the sciences studying them to have their existence dependent on their inherence in a more fundamental subject, which is an epistemic subject. On the basis of evidence within the *Posterior Analytics* as well as within the more properly metaphysical writings of Aristotle, I show how Aristotle would consider an epistemic substance as not having its existence explained on the basis of the essence of any other kind.

Within the first book of the *Posterior Analytics* Aristotle makes a number of observations concerning those first principles of the sciences that are assumptions of existence and concerning the relation they have to those first principles that are definitions. Aristotle's remarks here are disjointed and even more crabbed than usual, and, as a consequence, a scholarly consensus concerning Aristotle's views on assumptions of existence has not been reached. In chapter 3 I apply to these texts the results of chapter 2. I argue that the epistemic substances are those entities the existence of which must be assumed by the sciences studying them, through what Aristotle calls "hypotheses," and that the definitions of such entities, and no other, will have the status of scientific first principles. I then argue against two important heterodox accounts of the nature of assumptions of being (assumptions of τὸ εἶναι τι and of εἰ ἔστι) as accounted for in the *Posterior Analytics*, those of Gomez-Lobo and Hintikka. I close the chapter by reintroducing the problem of how it is possible to demonstrate καθ' αὑτὰ συμβεβηκότα predications. The καθ' αὑτὰ συμβεβηκότα predicates are those entities considered by the sciences, that are not epistemically substantial. The demonstrative explanation of why such a predicate exists (as inherent in some subject) cannot rest on a first principle that is a definition of the predicate, for in the case of such entities there are no definitional first principles. Nevertheless, such a demonstration requires *some* kind of definitional account, by virtue of which one is able to fix the reference of the καθ' αὑτὸ συμβεβηκός predicate. I suggest that it is *this* sort of definition that can, in a sense, be demonstrated.

Posterior Analytics 3–7 is a set of aporetic chapters, in which Aristotle presents the conceptual problems that an adequate account of a demonstration of a definition must meet. As is Aristotle's usual practice, the dialectical discussion of these problems brings to light insights that make possible a solution to the main problem at hand. In chapter 4 I comment upon the main points of this aporetic discussion.

I begin chapter 5 by presenting and commenting upon several versions of the prevailing line of interpretation of *Posterior Analytics* 2.8–9, all of which have their source in the commentary of Philoponus. According to this line of interpretation, when Aristotle shows how there can be a demonstration of a definition, he is, in fact, presenting an account of how there can be a demonstrative explanation of that kind of definition that can function as a first principle of the sciences. As so understood, Aristotle shows how one aspect of that definition (the "material definition") follows logically from another aspect of that definition (the "formal definition"). I show how this line of interpretation cannot form the basis of an adequate account of the text.

There follows the heart of the book, a close commentary of *Posterior Analytics* 2.8–9, which builds on the results of the previous four chapters. Following Themistius and Ross, I argue that Aristotle there is concerned with the "demonstration of the definition" of only those entities that are not epistemically substantial, that is, those that are καθ' αὑτὰ συμβεβηκότα. My thesis is that the demonstration of the definition of a καθ' αὑτὸ συμβεβηκός predicate is identical to the demonstration of the inherence of that predicate in some epistemic substance. Aristotle shows how such a demonstration must rest on a *nominal* definition of the καθ' αὑτὸ συμβεβηκός predicate, that is, a definition that fixes the conceptual content or meaning of the term that denotes that predicate. Unlike the definitions employed by the sciences, such a definition will have its sanction in ordinary, unscientific usage. If one can show that the predicates that make up the meaning of the term as so defined are in fact exemplified as inherent in some epistemic substance, one has in fact scientifically demonstrated that the καθ' αὑτὸ συμβεβηκός predicate, as conceived in a nonscientific manner, is in fact predicated of that epistemic substance. In so doing, one has also, in a sense, scientifically *explained* the nominal definition of that predicate, though this demonstration is not a syllogism whose conclusion is that nominal definition as such. It is in this sense, and this sense only, that Aristotle believes that a demonstration of a definition is possible.

I conclude chapter 5 with a close reading of *Posterior Analytics* 2.10, which I take to be a recapitulation of the preceding chapters of book 2.

In chapter 6 I explore three repercussions of the interpretation that I outline. First, I show how it allows one to solve the problem concerning the possibility of the demonstration of καθ' αὑτὰ συμβεβηκότα predications, to which Ferejohn and McKirahan have brought our attention. In short, the solution is as follows. In demonstrating that S is P, when P is a καθ' αὑτὸ συμβεβηκός predicate, one explains why there is predicated of S all of those predicates that are contained in the nominal definition of P. This is done by considering S from the perspective of more than one of the sciences. Second, I show how such a demonstration, as so understood, calls into question the extent to which Aristotle's philosophy of science commits him to the thesis of the autonomy of the sciences. This is a surprising result but, in my view, is an unavoidable consequence of Aristotle's investigations within the second book of the *Posterior Analytics*. Third, I show how my account of the logical structure of demonstrations of καθ' αὑτὰ συμβεβηκότα predications can be confirmed by scientific explanations that Aristotle actually presents, within the *Posterior Analytics* and elsewhere.

2

Aristotle on the Objects of Explanation

It is now generally recognized that Aristotle's main concern in the *Posterior Analytics* is the explication and articulation of the logical structure of scientific explanations, not the establishment of a program or methodology for scientific research and inquiry.[1] Thus, within the first book Aristotle shows how one scientifically explains certain facts and phenomena concerning certain kinds by showing how they logically follow from certain first principles. Among these principles are definitions, which express the essences of those kinds. But, as Aristotle elsewhere makes clear (*Phys.* 1.1), the definition of a kind may well be an insight gained at a late stage in one's research. Accordingly, the first book of the *Posterior Analytics* explores the inner structure of a body of knowledge already completed; it does not discuss the steps that are involved in gaining this knowledge in the first place.

When we turn to *Posterior Analytics* 2.1–2, however, we find that Aristotle's concern has abruptly shifted. In these chapters Aristotle distinguishes four kinds of inquiry and determines the *temporal* (as opposed to logical) relations that hold among them. Then in the following chapters (2.3–10), in which Aristotle sets before himself and subsequently solves the problem concerning the possibility and nature of demonstrations of definitions, there is again an apparent shift to a concern with the logic of scientific explanations already attained. Here Aristotle appears to revert from a consideration of the way in which one arrives at scientific explanations to considering the *logic* of scientific explanations already achieved, more particularly, the logic of a "demonstration of a definition." Are 2.1–2 anomalies in the *Posterior Analytics* as a whole?

I propose that they are not. A close reading of these chapters reveals that they too have as a primary purpose the clarification of the logical

1. This is the key contribution that Burnyeat has made to understanding the *APo.* ("Aristotle on Understanding Knowledge," in *Aristotle on Science: Proceedings of the Eighth Symposium Aristotelicum*, ed. E. Berti [Padua: Editrice Antenore, 1981]).

relationships that hold among those propositions that are the objects of scientific understanding. Further, Aristotle's main interest in 2.8–9, the sort of demonstration that explains a definition, is such that one can appreciate its logical structure only if he or she appreciates the temporal sequence of steps one follows in arriving at this demonstration. This is because, as I will show in chapter 5, this demonstration proceeds by showing how there is exemplified the conjunction of those predicates that, prior to gaining scientific understanding of the entity to be defined, one took to constitute the meaning of the term referring to this entity. So to account for the structure of the "demonstration of the definition" of this entity, Aristotle must account for the sequence of partial understandings of this entity that precede the scientist's attaining this demonstration.

This is why *Posterior Analytics* 2.3–10 is preceded by 2.1–2, and why an account of the former must be preceded by my account of the latter. Accordingly, the task of the present chapter is to subject *Posterior Analytics* 2.1–2 to a close reading, with special attention paid to the questions that will be of prime concern to Aristotle when he gives his account of a "demonstration of a definition."

I shall show that one of Aristotle's main points is that all scientific inquiry into whether something is the case is at bottom inquiry into whether there is a scientific explanation for that fact and that a scientific inquiry into why something is the case is an inquiry into what those "middle terms" are that make an explanatory explanation possible. There are two reasons why this is important for the resolution of the problem of the logical form of the demonstrations of καθ' αὑτὰ συμβεβηκότα predications. First, this view, along with the theory of demonstration put forward in *Posterior Analytics* 1, leads Aristotle to exclude the theoretical primitives of the sciences and their definienda from those predicates whose existence can be explained by means of a standard demonstration. These are those things that at 2.8.93b19 and 9.23b21–25 are said to have "no different cause" and therefore cannot have their definitions "demonstrated." This is a crucial step in my argument that within 2.8 and 2.9 Aristotle is giving an account of the logical form of demonstrations of καθ' ἁυτὰ συμβεβηκότα predications. Second, the insight expressed in 2.1 that any determination of essence must be preceded by a sufficient determination of existence provides for Aristotle the key to solving the problem of how one can arrive at an explanatory demonstration of a καθ' αὑτὸ συμβεβηκός predicate's inherence in its subject: one begins with that conception of the predicate that one holds prior to the scientific investigation of the predicate and then, on the basis of the first principles of the sciences, explains the existence of the predicate, as so conceived.

A. The Objects of Inquiry: *Posterior Analytics* 2.1–2

Aristotle begins 2.1 by noting that the kinds of objects of inquiry are equal in number to the kinds of objects of scientific understanding. These are τὸ ὅτι (the fact that . . .), τὸ διότι (the reason why), εἰ ἔστι (if it is), and τί ἐστιν (what is it) (2.1.89b24–25). Examples Aristotle provides of these are, respectively, that the Sun is eclipsed, why the Sun is eclipsed, whether a centaur or a god is (or is not), and what a god or human is (89b26–35). These examples allow us to infer Aristotle's meaning with a fair degree of certainty. By the inquiry into τὸ ὅτι Aristotle has in mind that which investigates whether there holds some *predication* (that some subject S has predicated of it some predicate P). By the inquiry into τὸ διότι he has in mind that which investigates *why* it is the case that such a predication holds (why S is P). The inquiry into εἰ ἔστι is that which investigates whether some subject exists (whether it is the case that something is S). The inquiry into τί ἐστιν is that which investigates the essence of some subject (what it is to be S).[2]

Within this short chapter Aristotle also remarks that an inquiry into the reason why a fact is the case (τὸ διότι) must temporally follow a determination that the fact (τὸ ὅτι) is indeed the case. Presumably, this is because ascertaining the fact ensures that there is indeed an object to the inquiry into why the fact is the case. (To employ one of Aristotle's examples, we know that the inquiry into the cause of the eclipse of

2. See W.D. Ross, *Aristotle's "Prior and Posterior Analytics"* (Oxford: Clarendon Press, 1949), 609–10; J. Barnes, *Aristotle's "Posterior Analytics," Translated with Notes* (Oxford: Clarendon Press, 1975), 194–95; and R. McKirahan *Principles and Proofs: Aristotle's Theory of Demonstrative Science* (Princeton: Princeton University Press, 1992), 188–89. A heterodox interpretation of εἰ ἔστι here has been presented by A. Gomez-Lobo ("The So-Called Question of Existence in Aristotle, *An. Post.* 2.1–2," *Review of Metaphysics* 34 [1980]: 71–89). Gomez-Lobo there argues that to inquire whether εἰ ἔστι θεός is to ask "is it a god?" where the understood subject is some particular thing understood by the context. Gomez-Lobo's treatment of this issue is closely related to his treatment (in "Aristotle's Hypotheses and the Euclidean Postulates," *Review of Metaphysics* 30 [1977]: 430–39) of Aristotle's positing assumptions of τὸ εἶναί τι ἢ τὸ μὴ εἶναί τι at 1.2.78a18–20 as assumptions that particular things are of particular kinds. Gomez-Lobo's views are addressed in chap. 3, sec. C.

some celestial body has some object only after we have ascertained that the celestial body does indeed undergo eclipse.) Aristotle similarly asserts that an inquiry into a kind's essence (τί ἐστιν) must be preceded by a positive resolution of the inquiry into whether there is such a thing as that kind (εἰ ἔστι). Here, too, this ensures that the inquiry has an object. Unless one knows that centaurs exist, there is no assurance that the inquiry into what centaurs are has any object at all.[3]

Aristotle begins the second chapter by relating this sketch of the kinds of inquiry to the theory of demonstrations:

> These are the things about which we inquire and, having discovered them, we know, and they are of this number. When we inquire about what is the case (τὸ ὅτι) or if something is in an absolute sense (εἰ ἔστιν ἁπλῶς), we inquire about whether there is or is not a middle term for it. And when we know what is the case (τὸ ὅτι) or if something is (either partially or in an absolute sense) (εἰ ἔστι, ἢ τὸ ἐπὶ μέρους ἢ τὸ ἁπλῶς), we further inquire about the reason why it is the case or the "what is it" (τὸ διὰ τί . . . ἢ τὸ τί ἐστιν); at this point we inquire about what the middle term is. (2.2.89b36–90a1)

Aristotle here distinguishes partial being and being in an absolute sense. For S to "be partially" is for it to have some predicate P predicated of it; for S to be "in an absolute sense" is for S to exist at all. Examples of the former kinds of being are "the Moon is eclipsed" and "the Moon waxes"; examples of the latter are "the Moon exists" and "night exists" (2.2.90a2–3). Here, as usual, Aristotle's terminology is somewhat fluid. The being of a state of affairs, which in 2.1 was called τὸ ὅτι, is now called being ἐπὶ μέρους, while the existence of something, which in 2.1 was called εἰ ἔστι, is now called being ἁπλῶς.[4]

---

3. More needs to be said on this issue. For it could be plausibly argued that quite a bit can be known about nonexistent objects: we know that centaurs are animals, etc. On this issue, see chap. 3, sec. E.

4. See Ross, *Aristotle's "Prior and Posterior Analytics,"* 612; McKirahan, *Principles and Proofs*, 190; and Barnes, *Aristotle's "Posterior Analytics,"* 194, who cites as a parallel *SE* 1.5.166b37–167a4. Much of Deslauriers' treatment of 2.2 is vitiated by the error of taking that which is ἐπὶ μέρους to be the predicate that is said to belong to a subject ("Aristotle's Four Types of Definition," *Apeiron* 23 [1990]: 3, 5). The equation of εἶναι ἐπὶ μέρους with εἶναι τοδὶ ἢ τοδί at 90a9–10 indicates that the phrase ἐπὶ μέρους has the meaning of "in

A few lines later the distinction is made in yet another way:

> Is it eclipsed? Is there some cause or not? After these considerations, knowing that there is some cause, we inquire what this is. For the cause of its being not this or that but of its being in an absolute sense in respect to its essence (εἶναι ἁπλῶς τὴν οὐσίαν),[5] or of being not in an absolute sense but of being one of the in-respect-to-itself (καθ' αὑτό) or accidental predicates is the middle term. I mean by [that which is] in an absolute sense the subject (ἁπλῶς τὸ ὑποκείμενον),[6] for example, a moon or earth or sun or triangle, and by one of the attributes I mean an eclipse, equality, inequality, and whether it is in the middle or not. For in all of these cases the "what is it" and the reason why are the same. (2.2.90a7–15)

The distinction between a subject S being P (in its having P predicated of it) and there being S at all is now referred to as that between being "this or that" or being in respect to "some in-respect-to-itself or accidental predicate," on the one hand, and being "in an absolute sense in respect to its essence" or being "in an absolute sense the subject," on the other hand. Aristotle employs τὸ ὅτι ἔστι (89b37, 39; 90a2, 8), εἰ ἔστι (89b38, 39; 90a3, 4), and εἶναι (90a9) synonymously, showing that he is

---

a certain respect" and that for a subject to be in a certain respect is for it to "be something or other," for it to have something or other predicated of it. The examples at 13–14 (eclipse, equality, inequality, in the middle or not) are not examples of what is ἐπὶ μέρους but, rather, of the possible values of τοδί, the possible predicates of a subject.

5. Contra Ross (*Aristotle's "Prior and Posterior Analytics,"* 611), the term οὐσία cannot here mean "substance," since one of the examples Aristotle gives of "being in respect to the οὐσία is "triangle" (90a13), which is neither a substance nor considered as a substance by the geometer. A survey of the use of οὐσία in *APo.* reveals only one passage (1.4.73b7–8) in which it clearly has the sense of "substance." At 1.27.79a7 and 1.22.83a24–b12 οὐσία is apparently used to refer to what in chapter 3 I label "epistemic substances." In at least thirteen passages οὐσία means "essence": 1.4.73a36; 1.13.79a6–7 (see J. Tricot, *L'Organon d'Aristote, traduction nouvelle et notes,* 2d ed. [Paris: J. Vrin, 1979], 78; and J. Pacius, *In Porphyrii Isagogen et Aristotelis Organum, Commentarius Analyticus* [Hildesheim, Ger.: Olms, 1966], 311–12); 1.32.89a19–21; 2.3.90b16–17; 2.4.91b7–9; 2.6.92a6; 2.7.92a34–35, 92b12–14, 92b29; 2.9.93b25–28; 2.13.96a34–35, 96b6, and 97a12–14.

6. The example "triangle" at 90a13 shows that Aristotle is here using the term ὑποκείμενον in its grammatical sense as the subject of a proposition, not in its metaphysical sense as a real substrate, or even in its sense as the ultimate substrate posited by a given science (for geometry considers a triangle an attribute of a group of either points or lines [on this, see chap. 3, sec. B]).

no longer following the convention established in the previous chapter.[7]

In this passage Aristotle is claiming that the inquiry concerning whether something is the case (whether some S is P) or whether some subject S exists is also an inquiry into whether or not there is some middle term that would make possible a syllogism whose conclusion would be the predication or assertion of existence at issue. This middle term, which is identified as the cause of such a conclusion, is also identified as what something is. It is an essence.

A number of problems arise here. First, the inquiry into whether something is the case or exists can be identified with the inquiry into whether a state of affairs or fact has a cause only if it is denied that there are uncaused states of affairs or existents. Is this a plausible assumption for Aristotle to make? Second, Aristotle leaves unclear exactly what he takes to be the middle term that identifies the cause of the object of inquiry. He tells us that it is an essence but not that of which it is the essence. Will it be the essence of the subject, of the predicate, or of something else? Third, what is the form of the explanatory syllogism Aristotle has in mind? An explanatory demonstration (an ἀπόδειξις τοῦ διότι), as described in *Posterior Analytics* 1.13.78a39–b3, is a syllogism whose middle terms reveal why the predication that is its conclusion is the case; this is why it makes possible the ἕξις of scientific understanding by virtue of which we can explain why the predication is the case. But Aristotle tells us that, among other things, finding the right middle term will allow us to explain certain cases of existence as well as certain predications. How can an existence claim be the conclusion of such a demonstration?

The first question can be answered in one of two ways. First, one can plausibly deny that Aristotle believes that there are uncaused events and existents. Although Aristotle grants that some occurrences and attributes are "accidents," or "coincidences," (τὰ συμβεβηκότα), it can be argued that these are to be considered uncaused only under a certain description, that is, when considered as inhering in or occurring to a substrate to which there is no direct causal link.[8] Alternatively, it can be

7. On this, see Ross, *Aristotle's "Prior and Posterior Analytics,"* 610; Barnes, *Aristotle's "Posterior Analytics,"* 195; and T. Upton, "The If-It-Is Question in Aristotle," *Ancient Philosophy* 11 (1991): 316. Aristotle similarly employs ὅτι ἔστι to refer to a thing's existence at *APo.* 1.1.71a16. On this passage see chap. 3, sec. B.

8. See *APo.* 1.4.73b10–16, and *Metaph.* 5.30.1025a14–27.

argued that Aristotle is restricting himself to a discussion of those inquiries whose objects are objects of the sciences and are, accordingly, such as have a cause which can be identified by one with the appropriate scientific understanding (cf. *APo.* 1.2.71b9–12). This is supported by the fact that, as Barnes has pointed out, 2.2 is manifestly a continuation of 2.1, which opens with the statement that the objects of inquiry are equal in number to the objects of scientific understanding.[9] The latter alternative seems to me more plausible, but, for my purpose here, either account will do.

The other questions raised by Aristotle's identification of the middle term that reveals a cause and an essence are in part clarified by the next few lines, in which Aristotle presents some examples:

> What is an eclipse? A privation of light from the moon on account of the blocking of light by the earth. Why is there an eclipse, or rather why is the moon eclipsed? Because there is a lack of light when the earth acts as a screen. What is a harmony? A numerical ratio between high and low. Why does the high harmonize with the low? Because the high and low have the numerical ratio. Do the high and low harmonize? Is there a numerical ratio between them? Given that there is, what then is the ratio? (2.2.90a15–23)

So what is meant by the identification at 90a14–15 of the "why is it?" (διὰ τί ἔστιν) question with that asking "what is it?" (τὸ τί ἐστι)? Two reasons why this is unclear are the fact that the ἔστι in the first question can have more than one sense and the fact that the subject of the second question is unspecified. Aristotle could be saying that we answer a "why is the subject S a P?" question by identifying the essence of S. In this case the ἐστι[10] in the question διὰ τί ἐστι would be the copula, a case of what had been earlier called partial being (εἶναι ἐπὶ μέρους), being this or that (εἶναι τοδὶ ἢ τοδὶ), and being not in an absolute sense

9. Barnes, *Aristotle's "Posterior Analytics,"* 195. I do take issue with Barnes' suggestion that Aristotle is here discussing only inquiries into mediated predications. On this, see sec. B.

10. I alter the accentuation in accordance with that mandated by the interpretation being discussed. In most of the relevant texts Aristotle omits some form of εἶναι as understood, and, of course, even in those cases in which he does not, there is no evidence of what Aristotle's own accentuation is. Further, most scholars today maintain that the accentuation of forms of εἶναι was not determined by the sense of the verb. See C. Kahn, *The Verb "Be" in Ancient Greek* (Dordrecht, Neth.: D. Reidel, 1973), 420–24.

but being one of the in-respect-to-itself or accidental predicates (εἶναι μὴ ἁπλῶς ἀλλά τι τῶν καθ' αὑτὸ ἢ κατὰ συμβεβηκός) (89b39–90a3, 90a9–11), and the subject of the question τί ἐστι would refer to the subject S. Second, the ἐστι in the question διὰ τί ἐστι could again be taken to be the copula, but the subject of the question τί ἐστι could be taken to refer to the predicate P. In this case Aristotle would be saying that we answer a "why is S a P?" question by identifying the essence of P. Third, taking the ἔστι in the expression of inquiry into the διὰ τί ἔστι to be existential, Aristotle could be saying that the answer to a question of the form "why does P exist?" is to be found in the essence of P.[11]

There is textual support for each of these possibilities. First, it should be noted that there are two possible cases in which one would ask whether a P exists. P can be considered a subject. In this case the inquirer would be asking whether P is "in an absolute sense the subject" (ἁπλῶς τὸ ὑποκείμενον [2.2.90a12]). But ancient Greek allows one to refer to a subject as "the P" where the subject is not P in respect to its own nature. Per Aristotelian language in such a case P is not the essence of the subject but is one of its accidents (cf. *APo.* 1.4.73b5–10). In some cases, such as τὸ λευκόν ἐστι ξύλον (the white [thing] is a log), it is apparent that the predication is equivalent to another predication (in this case, "the log is white") whose grammatical structure mirrors the metaphysical structure of the state of affairs it expresses. Such a predication will be one whose subject is substantial, the substrate in which the predicate truly inheres (cf. *APo.* 1.22.83a1–23). An inquiry into whether white exists (εἰ ἔστι τὸ λευκόν) will be the same as an inquiry into whether it is the case (ὅτι ἐστι) that something is white.[12] In such a case the second and third of these alternative interpretations of Aristotle's identification are to be identified. To ask why P exists is the same as asking why S is P, and the answer is to be found in the essence of P. To ask why there is the color white (considering the color white as a subject) is to ask why the color white inheres in certain surfaces (considering the color white as a predicate); the answer is to be found in determining what it really is for a surface to be white. Once we show why the essence of the color white is predicated of some surfaces, it is

11. See Barnes, *Aristotle's "Posterior Analytics,"* 196.

12. See G.E.L. Owen, "Aristotle on the Snares of Ontology," in *New Essays on Plato and Aristotle*, ed. R. Bambrough (New York: Humanities Press, 1965), 77–78, 87; reprinted in *Logic, Science, and Dialectic*, ed. M.C. Nussbaum (Ithaca, N.Y.: Cornell University Press, 1986); McKirahan, *Principles and Proofs*, 38, 196.

an easy task to show (through a first-figure syllogism) why the color white itself is predicated of those surfaces.

Among the predicates that admit nominalization while not being substantial are those that Ross aptly calls events;[13] these are temporally localized relations between subjects. Examples are night, an eclipse, and thunder. We are apt to make existential claims of such predicates (e.g., "there is thunder," and "eclipses exist"). Further, except for those who have mastered scientific explanations of these events, their metaphysical subjects are not as clear as are the subjects of predicates such as "white." This is because events are less persistent and are constituted by more than one substance standing in a certain relation. In such a case it is not immediately clear how to translate the inquiry into the predicate's existence into an inquiry into whether the predicate inheres in some particular subject. It is not insignificant that the central examples Aristotle presents in 2.2 (as well as those on which he focuses in 2.8 and 2.10) are of this kind. Aristotle indicates that to ask why the lunar eclipse exists (διὰ τί ἔστιν ἔκλειψις) is to ask why the Moon undergoes the lunar eclipse (διὰ τί ἐκλείπει ἡ σελήνη); this is to be resolved by an expression of the essence of the lunar eclipse (τί ἐστιν ἔκλειψις): a blocking of the light of the Moon through the interposition of Earth (2.2.90a15–17; cf. 2.8.93b2–3). (More precisely, since the Moon is not always eclipsed, the inquiry will concern why the Moon *sometimes* undergoes eclipse, and this is to be resolved by determining just what it is to be subject to eclipse. Clearly, this answer will at least include the answer to the question of what a lunar eclipse is.) Similarly, to ask whether there is such a thing as harmony is to ask whether it is possible for high and low notes to harmonize (ἆρα ἔστι συμφωνεῖν τὸ ὀξὺ καὶ τὸ βαρύ;); to understand why the high and low harmonize (διὰ τί συμφωνεῖ τὸ ὀξὺ τῷ βαρεῖ) one must need to know the essence of the predicate "harmonize": namely, to have a certain numerical ratio (2.2.90a18–23).

These examples show that in some cases Aristotle means to identify the answer to the "why is S a P" question with a statement of what P is. But Aristotle would also endorse the first alternative interpretation of the identification of the inquiries into "what is it?" and "why is it?" This is the identification of the answer to the "why is S a P" question with a statement expressing the essence of S. The reader of the first book of the

13. Ross, *Aristotle's "Prior and Posterior Analytics,"* 611–12.

Posterior Analytics would recall that the first principles or ultimate premises of explanatory demonstrations will be logical or general laws applicable to more than one genus (axioms), existence claims (hypotheses), and expressions of the essences of things (definitions) (see 1.2.72a14–24; 10.76a31–36, 76b3–16).[14] Hence, an expression of an essence will play a double role for Aristotle; it will ground the explanation of certain attributes of that thing as well as express its ontological constitution. The syllogistic form that Aristotle holds demonstrations must take (the first-figure universal syllogism, which the medieval logicians called "Barbara") necessitates that a definitional predication that is employed as a basic premise in demonstrating some predication S is P must be definitional of S or of one of the "middle terms"; it is impossible for one of these basic premises to be definitional of P.[15]

Let us reconsider the case in which the inquiry into the existence of a predicate P is the inquiry into the existence of what is properly considered as inherent in a subject S, the scientific inquiry into why S is P, or, in other words, why P exists. The first book of the *Posterior Analytics* indicates that such an inquiry is successfully resolved by determining the essence of S, which will be a middle term of the demonstration explaining why S is P. On the other hand, the examples Aristotle employs to elucidate the relationship between essence and explanation of existence indicate that such an inquiry is successfully resolved by determining the essence of P and that it is *this* that will serve as a middle term of the demonstration explaining why S is P. Can these accounts be reconciled? Presumably, all but the simplest demonstrations will have more than one middle term. But, within the framework of Aristotelian syllogistic, it is hard to see how both the essence of S and the essence of P can serve as middle terms. There will need to be a

14. Aristotle's discussion on the various kinds of first principles is notoriously unclear, but the interpretation presented here, which is the traditional one, seems to best account for most of what Aristotle says. I explore some of the complexities of this text in chap. 3, sec. A.

15. For example, the demonstration explaining why vines shed their leaves will have as premises "all vines are flat leaved," "all flat-leaved plants have the sap coagulating at the juncture of the leaf and the stem," and "everything having the sap coagulating at the juncture of the leaf and the stem sheds" (*APo.* 2.16.98b5–10, 17.99a28–29). The latter two premises are not definitional but, presumably, are demonstrable on the basis of definitional predications. The last such premise cannot be definitional of "sheds"; *sheds*, or some synonymous term, will be definitional of some implicit middle term. On what this middle term is, see chap. 7, sec. B.

premise in which the essence of P is predicated of the essence of S. If the essence of S is to be expressed as the conjunction of the genus and the differentia of S, how is this possible? For example, given a definition of the Moon as a celestial body of a certain kind, on what grounds will one be scientifically justified in predicating of that essence of the Moon whatever it is that we take to be the essence of the attribute of being subject to a lunar eclipse? Such a premise would itself need to be demonstrated, and it is hard to see how such a demonstration could be forthcoming. Certainly, the essence of being subject to eclipse is not implicitly definitional of the Moon.

It should be clear to the reader that we again face the problem of the possibility of the predication of καθ' αὑτὰ συμβεβηκότα predications, the elucidation of which was the central concern of chapter 1. Here we must again leave this as a yet unsolved problem.

B. How Are Unmediated Existence Claims to Be Explained?

One important possibility has not been considered in this discussion of the possible objects of scientific inquiry. Suppose that there is an εἰ ἔστι (whether it exists) inquiry in regard to a subject S that is not to be considered as inherent in another subject. For example, suppose that a science were to investigate the existence of a substantial kind, that is, whether lemurs exist. How will it be the case, as Aristotle at 2.2.90a5–11 says it is, that the inquiry will be into whether there is a *middle term* expressing the cause of the existence of S? For any demonstration explaining the existence of S would have to take one of two forms: a demonstration having as its conclusion "S′ is S" (e.g., "Every S′ is a lemur") where S′ is another subject in which S is shown to inhere, or one having as its conclusion "S exists" (e.g., "lemurs exist"). Neither is possible. For *ex hypothesi* there is no substrate S′ in which S inheres. And, even if we leave aside the issue of the status of the predicate "exists," it is evident that any syllogism whose conclusion would be of the form "S exists" would necessarily rest on a major premise "S is P," in which the existence of the major term S is already presupposed.[16]

16. I hold the orthodox opinion that Aristotle is committed to the view that the truth of a predication presupposes the existence of its subject. On this, see chap. 3, sec. E, in which I argue that it is this view that is responsible for Aristotle's requirement that there be assumptions of existence for all of the subjects of the sciences.

Both Ross and Barnes are aware of this problem. As we saw earlier, Barnes notes that the present passage is apparently concerned with only those inquiries that are concerned with the objects of science; Barnes suggests that this is meant to exclude inquiries into indemonstrable propositions.[17] It is true that Aristotle often distinguishes intellect (νοῦς), the disposition (ἕξις) by which we grasp the first principles of the sciences, from science, the disposition by which we can provide demonstrative explanations of mediated propositions (*EN* 6.6; *APo.* 1.3.72b23–25, 2.19.100b5–15). But Aristotle's citation at 90a5 of "does the Moon exist?" as an example of the sort of inquiry he has in mind is reason enough to reject Barnes' suggestion. For the Moon is a substance, which does not inhere in any more ultimate subject. Hence, the existence of the Moon cannot be demonstrated by explaining its inherence in some more basic subject.

Ross's appraisal of the problem is worthy of quotation:

> But how can εἰ ἔστι or τί ἐστι applied to a *substance* be supposed to be concerned with a middle term? A substance does not inhere in anything; there are no two terms between which a middle term is to be found. A. gives no example of what he means by the μέσον in such a case, and in this chapter the application of the questions εἰ ἔστι and τί ἐστι to substances is overshadowed by its application to attributes and events, which is amply illustrated (90a15–23). He does not seem to have thought out the implications of his view where it is the εἰ ἔστι or τί ἐστι of a substance that is in question, and the only clue we have to his meaning is the statement that by μέσον he means αἴτιον. As regards the εἰ ἔστι of substances, then, he will be saying that since they, no less than attributes, must have a certain ground of their being, to inquire whether a certain substance exists is by implication to inquire whether something that is its cause exists. As regards the τί ἐστι of substances he will be saying that to inquire into what a certain substance is, is to inquire what its cause is; i.e. that its definition, no less than that of an attribute, should be causal, that a substance should be defined by reference either to a final or an efficient

17. This position is attributed to Aristotle by Barnes (*Aristotle's "Posterior Analytics,"* 195) and McKirahan (*Principles and Proofs*, 190).

> cause. This is the doctrine laid down in *Met.* 1041a26 . . . But it cannot be said that A. remains faithful to this view . . .[18]

Ross recognizes that what Aristotle means by identifying the cause of a thing's being and its μέσον is not at all evident when the thing concerned is a substance. To clarify what Aristotle *should* have said, Ross turns to *Metaphysics* 7.17.1041a26–32, in which Aristotle investigates in the light of his mature metaphysics the very question of the structure of the explanation of the existence of a substance. In this passage Aristotle takes the cause of the existence of a substantial kind to be the essence of that kind itself, here identified with the final or efficient cause of that kind. One might think that it would be natural to employ this passage in support of the view that substantial kinds are among the kinds that have "no different cause" (mentioned at *APo.* 2.8.90a5–6, 93b19; and 2.9.93b21), a view advocated by both Ross and myself. But, when he discusses what Aristotle in the *Posterior Analytics* means by "what has no other cause," Ross does not support his position by referring to *Metaphysics* 7.17.1041a26–32. There are three possible reasons for this. First, the passage from *Metaphysics* is itself far from clear; in particular, Aristotle's identification of the essence of a kind with its final and efficient causes is quite obscure. Second, the argument of the whole of *Metaphysics* 7.17 rests heavily on the metaphysical distinction between form and matter, which is entirely absent from the *Organon*. Third, as Ross himself points out just following the preceding passage, within the *Organon* the canonical form for the definition of a kind is its genus conjoined with its differentia. It is unclear how such a definition, whose aim is apparently *classificatory*, can fill the role of *explaining* why there is such a kind.

Nonetheless, *Metaphysics* 7.17.1041a26–32 does provide some support for the interpretation of "that of which there is no different cause" in *Posterior Analytics* 2 for which I shall argue. Within *Metaphysics* 7.17 Aristotle employs the principles of the theory of explanation expounded in the *Posterior Analytics* in order to shed light on metaphysical questions. Starting from the account of *Posterior Analytics* on the kinds of scientific inquiry and the nature of their objects, Aristotle shows how metaphysical explanations posit substances as principles and causes (*Metaph.* 7.17.1041a6–10, b8–9; 27–31). In doing so, Aristotle

18. Ross, *Aristotle's "Prior and Posterior Analytics,"* 612.

explicitly addresses the question we have seen on the horizon of *Posterior Analytics* 2.1–2: What is the object of the inquiry into those kinds that are not considered as inherent in any other subject? What is the cause of the existence of such a kind? Now we can expect that, unless there is clear evidence to the contrary, Aristotle's views in *Metaphysics* 7.17 on the structure of explanations of existence are consonant with the views on this subject put forward in the *Posterior Analytics*. Though we cannot impute the metaphysical views of the later work to the *Posterior Analytics*, *Metaphysics* 7.17 can shed light on Aristotle's views concerning the *nonmetaphysical, scientific* explanation of the existence of a basic kind.

C. *Metaphysics* 7.17 on the Explanation of Substantial Existence

In the last section I considered Aristotle's assertion in *Posterior Analytics* 2.2 that a demonstration of the existence of a being will explain that existence by means of a middle term, which serves to identify the cause of that existence. Following Ross, I have shown how this is problematical in the case in which the being whose existence is to be explained is not considered as inherent in any more fundamental subject. The only text in which Aristotle addresses this issue is *Metaphysics* 7.17, which I here examine. I shall conclude that Aristotle holds the (apparently paradoxical) position that the cause of the existence of any being not inherent in any more fundamental kind is that kind itself. In *Metaphysics* 17 this position is explained by appeal to principles of Aristotle's mature metaphysics, which are absent from the *Posterior Analytics*. I speculate on how this position would be understood within the context of the *Posterior Analytics*. These speculations will be provisional until they can be supported by employing them in my account of *Posterior Analytics* 2.8–10.

Aristotle begins *Metaphysics* 7.17 by asserting that every inquiry (ζήτησις) is an inquiry into the predication of one thing of another. Thus, to ask why thunder exists is to ask why noise is in the clouds (1041a10–14, 23–26). This is consonant with *Posterior Analytics* 22; to ask about the cause of the existence of something is to seek the middle term that explains why that thing inheres in something else. We have seen that this strategy will not work for substantial predicates such as "human"; to ask why there are human beings is not to ask why the predicate "human" inheres in some more basic subject. Aristotle here addresses the problem by taking up the example of the inquiry into the existence of human beings. Instead of calling this the inquiry into why human beings exist (διὰ τί ἄνθρωπος ἔστι), he calls this the inquiry into why a human being is a human being (διὰ τί ὁ ἄνθρωπος ἄνθρωπος). According to the latter formulation the inquiry seems to be into the cause of a tautology, and this is indeed Aristotle's original response. But, as Aristotle's argument progresses, it becomes clear that it is the very existence of a human being that is the real subject of inquiry.

Within *Metaphysics* 7.17 Aristotle's initial response to the problem of why a human being is a human being is to cut the Gordian knot: such an inquiry is without an object. The inquiry would have an object only if we take it to be an investigation of the logical relations that hold between unity and identity. Such an inquiry would be easily resolved and would pertain to all beings, not human beings as such (1041a16–22).

This solution is, of course, unsatisfactory. People obviously do inquire into whether certain ultimate substrates exist, considered as being of a certain kind, as, for example, when they investigate whether there is a Loch Ness monster. We have seen that Aristotle holds that, if it is determined that a certain kind does exist, one seeks the cause of that existence. Aristotle does not explain how we pursue such an inquiry into why an ultimate substrate exists, merely by stipulating that the term *inquiry* is not to be employed in such a case. But a more adequate account of the matter is soon put forward.

This passage (in part cited by Ross in the passage from his earlier commentary) can be translated as follows:

> And why are these things, for example bricks and stones, a house? It is clear that one is seeking the cause. And this is the essence (τὸ τί ἦν εἶναι), speaking in respect to the formulae (ὡς εἰπεῖν λογικῶς),[19] which in some cases is that for the sake of which, say in the case of a house or a bed, and in other cases that which first moved it, for that too is a cause. But while the first kind of cause is sought (ζητεῖται) in the case of coming into being and passing

19. I follow Ross in including the phrase τοῦτο δ' ἐστὶ τὸ τί ἦν εἶναι, ὡς εἰπεῖν λογικῶς, which pseudo-Alexander and Jaeger excise. In giving λογικῶς the sense of "attending to formulae," I follow E. Halper (*One and Many in Aristotle's Metaphysics* [Columbus: Ohio State University Press, 1989], 43–44, 280–81 n. 181).

> away (γίγνεσθαι . . . καὶ φθείρεσθαι), it is also in the case of being (εἶναι) that the other is sought. (1041a26–32)

To understand why something exists is to grasp its cause. Aristotle explains that the cause that is sought will be the essence. But what is the essence? Aristotle says that this depends on the sort of thing whose essence is sought.

On the one hand, if the thing is something that *is*, the essence that is sought will be the final cause. As he often does, Aristotle presents examples of artifacts. To understand the being of a house is to understand why the building materials are arranged as they are: to provide the necessary shelter (cf. *De An.* 1.1.403b5–6). Likewise, to understand the being of a bed is to understand that the wood has been arranged a certain way in order to provide a surface on which to sleep. On the other hand, if one is inquiring why some attribute or thing comes into being or passes away, that is, why something *happens*, the efficient cause, as well as the final cause,[20] must enter into the essence of a thing.

Aristotle's distinction between εἶναι on the one hand and γίγνεσθαι and φθείρεσθαι on the other, so reminiscent of Platonic metaphysics, is obscure. Aristotle's account of the sort of causes that ought to be sought in each of the two cases is likewise unclear. I propose the following interpretation.

For Aristotle the world is composed of particular substances, but particulars can be considered either in respect to their particularity or insofar as they belong to some universal kind. Accordingly, there are three ways one can inquire why a human being is a human being. First, one can inquire why a particular human being is that particular human being. For example, one can inquire why Socrates is Socrates. I take it that it is this inquiry that Aristotle takes to be futile, answerable only by appeal to the principle that every unity is identical to itself (1041a16–20). Second, one can inquire why a particular belongs to the kind "human being." For example, one can inquire why Socrates is human. Third, one can inquire why the kind "human being" is what it is: the kind human being. This appears to be an inquiry into a tautology, as is the first inquiry. But we shall see that there is an important difference.

---

20. The καί at 1041a32 indicates that when *explanandum* is a case of coming into being, as well as when it is a case of being, the final cause is to be sought.

Let us first attend to the nature of the second kind of inquiry. Aristotle discusses this later in *Metaphysics* 7.17:

> Since one must know that existence belongs to the thing, clearly one is seeking why the matter is something. For example, why are these things a house? Because what it is to be a house is present. Likewise this thing here, or rather this body having that form, is a human. So that what is sought is the cause (and this is the form) by which it is something, and this is the essence (οὐσία). (*Metaph.* 7.17.1041b4–9)

We understand why a particular substance is the kind of thing it is by grasping the fact that its form is present in its matter. To understand why it is *this* bit of matter that has taken on a certain form is to inquire about the *coming into being* of the substance; to resolve this inquiry one would need to account for the history of the substance.[21] This is why Aristotle says that the essence that is sought in the case of a particular substance's coming to be involves the efficient cause and the final cause of that substance (1041a31–32). Aristotle has in mind his account of the metaphysics of generation, in both the natural and artificial realms. In order to understand the creation of an artifact, such as a bed, one must grasp that the efficient case, the carpenter, actualized the potentiality of the bed to take on the form of the bed, which is the essence of a bed and, consequently, the final cause of a bed. For the essence of a bed is to be a structure shaped for comfort and support in sleeping. Likewise, to understand the generation of a chicken, one must grasp that the efficient cause is a certain rooster, who in this case is the same in essence as the chicken in question, and that this rooster instilled in matter a certain form, the soul of the chicken, which is the first actuality to perform those life functions that are the final cause or purpose of that chicken itself. So we see why Aristotle indicates that the inquiry into why a particular belongs to a certain universal kind is an inquiry into the essence, efficient cause, and final cause of that particular.

It is possible, however, to inquire into particulars not in respect to their particularity but insofar as they fall under certain kinds. Instead of beginning by focusing on certain particulars, and inquiring into

21. See M. Frede, "Substance in Aristotle's Metaphysics," *Essays in Ancient Philosophy*, (Minneapolis: University of Minnesota Press, 1987), 78.

whether they are human beings—that is, whether the form being human is inherent in the matter—we can inquire whether this form itself exists. In this case there will be no need to seek the efficient causes of the fact that certain particulars fall under that kind.

It is tempting to identify the investigation of the existence of a universal kind with what Aristotle here calls an inquiry into the εἶναι of a thing. For such kinds can be identified with eternal species, which do not come to be and pass away. They are the concern of the sciences, which are concerned with truths that are necessary (*APo.* 1.2.71b9–16) and eternal (*APo.* 1.8.75b21–36). It would follow that at *Metaphysics* 7.17.1041a32 Aristotle is claiming that every essence of a universal kind includes the final cause of that kind but not its efficient cause.[22]

But there is a distinction between those kinds that are substantial and those that are not. Those that are substantial do not have their forms come into being (*Metaph.* 7.7.1033b5–8, 15.1039b23–27). An inquiry into why there exists such a substantial form *is* a case of inquiry into being, not an inquiry into a coming into being or passing away. We can interpret 1041a32 as making the point that this kind of inquiry involves seeking the essence correlated to that form, which includes the final cause of the form. It should be noted that the existence of such forms does not result from the fact that one thing inheres in another. For this reason at *Metaphysics* 7.17.1041a32, where it is said that in cases of *being* as opposed to becoming, the final cause is sought (ζητεῖται), Aristotle is apparently employing a more general sense of *inquiry* (ζήτησις) than that presented at 1041a10–11 and 25–26. This is an indication that Aristotle looks more kindly on the investigation of whether a certain form exists than he does on the investigation into whether a certain particular is itself. The latter investigation has no object; it seeks no new knowledge about the world. But to seek whether a certain form exists is an investigation with an object; one is seeking whether the world is the sort of place in which any particulars have a form of a certain kind. It is meaningful to ask whether there is a Loch Ness monster or whether there is a god (cf. *APo.* 2.1.89b32).

But not all inquiries into the existence of forms are such inquiries into the εἶναι of a thing. *Posterior Analytics* 2.8 and 10 make clear that natural but nonsubstantial phenomena or events, like the lunar eclipse,

22. This proposal is in essence that presented in M.F. Burnyeat, ed., *Notes on Book Zeta of Aristotle's "Metaphysics,"* Study Aids Monograph No.1. (Oxford: Subfaculty of Philosophy, 1979), 153.

for which there is no eternally existing form to be grasped by the intellect, are also to be considered as falling under universal kinds, to which there are correlated certain essences. One who wishes to discover such an essence seeks the reason why a certain natural phenomenon periodically occurs, that is, why some accidental attribute regularly comes to inhere in some substance. In this case what is sought is the efficient cause that is operative in all of these particular cases (*Metaph.* 8.4.1044b8–20).[23] The inquiries into the existence of such kinds, along with the inquiries into the existence of all particulars (whether substance or accident), will constitute all instances of what Aristotle at 1041a31–32 calls an inquiry into the cause of something's coming into being or passing away, an inquiry that according to Aristotle has as its aim the identification of the efficient cause.

Accordingly, at 1041a32 Aristotle asserts that, if one seeks the cause of the fact that a certain substantial kind exists, one seeks the essence of that kind, which will be its final cause. As *Metaphysics* 7.7.1032b1–2 makes clear, this essence will be the same as the form, which at 7.17.1041b4–9 is identified as a cause of the existence of particular material substances. This form does not itself come into being; it is not

23. In *APo.* 1.8.75b33–36, Aristotle explains how such cases do not call into question his assertion that the objects of science are eternal: "The demonstrations and sciences of the things that often come into being (τῶν πολλάκις γιγνομένον), such as of the eclipse of the Moon, are clearly eternal insofar as they are of a certain kind of thing (ᾗ . . . τοιοῦδ' εἰσίν), but insofar as they are not eternal, are partial (κατὰ μέρος)" (following manuscripts B and P). The passage is obscure, because it is not clear what the antecedent of τοιοῦδε is. The reading of the Greek that first suggests itself has τοιοῦδε refer back to the things that come about often. The meaning would then be that, although it is not the case that there are always lunar eclipses, a demonstration explains why a lunar eclipse, a thing of a certain kind, always has certain attributes. Barnes (*Aristotle's "Posterior Analytics,* 131) and McKirahan (*Principles and Proofs,* 128–32) present such an interpretation. This requires that Aristotle not consistently hold that science concerns eternally existing kinds; in chapter 3 I defend the traditional understanding that this is indeed Aristotle's considered position. But there is another alternative: τοιοῦδε could simply mean "of a certain kind of thing," leaving what kind unspecified. (Cf. Barnes' translation, "of such and such a thing" [14], which rendering does not commit one to the interpretation presented in his commentary). The point would then be that demonstrations concerning occasional events are eternal because they are really concerned with a certain kind of object, namely, an eternal object. For example, the demonstration that the Moon is occasionally eclipsed is a demonstration concerning the *subject* of the attribute in question, namely, the Moon, which subject *is* for Aristotle eternal. The fact that the Moon is occasionally eclipsed is an eternal fact, following from the eternal attributes of the eternal Moon. (Ross [*Aristotle's "Prior and Posterior Analytics,"* 532; cf. 63–64] may have this in mind in his paraphrase: "Demonstrations of things that happen often, in so far as they relate to a certain *type* of subject, are eternal.")

the result of one thing coming to be in something else (*Metaph.* 7.7.1033b5–8). Hence, when we ask about why this form or essence exists, we are not asking about why one thing comes to inhere in something else.

Aristotle's main concern in *Metaphysics* 7.17 is to show how οὐσία, in the sense of *essence*, is a primary cause. Whenever we inquire about why one thing belongs to something else, the answer is always "because of some form," which, as *Metaphysics* 7.7.1032b1–2 makes clear, is to be identified as the essence of the kind in question. If we give *inquiry* (ζήτησις) a special sense according to which it denotes inquiry into why one predicate belongs to some subject (as is advocated at *Metaph.* 7.17.1041a10–11), it follows from Aristotle's analysis that the cause of the existence of an essence or form cannot be the object of inquiry. I take it that this is at least in part what Aristotle means when he concludes his discussion of the cause of substantial being by noting, "It is clear then that there is neither inquiry nor teaching in the case of the simples (τῶν ἁπλῶν) but we deal with these things in a manner other than inquiry" (*Metaph.* 7.17.1041b9–11).[24] That is not to say that there is no ground in reality for the existence of such a kind, for to say this would be to say that the kind does not exist. The ground for the existence of the kind lies nowhere else but in the kind itself.

If *inquiry* is not given the restricted sense advocated at 1041a10–11, as it is not in *Posterior Analytics* 2.1–2,[25] then the middle term that is sought when one inquires about the cause of the existence of the essence of a substantial kind will be the final cause of that kind, which cause is the same as that essence itself.

There are, of course, major problems involved in casting into syllogistic form a scientific explanation of the existence of a substantial kind by taking its essence as a middle term. For the *explanans* of such a demonstration would be the same as the *explanandum*, and the demon-

24. One could also interpret this as saying that one cannot inquire into the cause of there being a simple subsistent form, such as an immaterial substance, because in this case the form/matter analysis is inapplicable. This may indeed be part of what Aristotle is saying. Nonetheless, such a statement would be odd in this context, since this would be the first time within *Metaph.* 7 that immaterial forms would be referred to as existent, not as beings whose existence is uncertain and needs to be investigated (*Metaph.* 7.2.1028b28–29). It seems to me more likely that by τὰ ἁπλά Aristotle is at least in part referring to forms of composite substances, considered apart from the matter in which they necessarily inhere.

25. At *Metaph.* 7.17.1041a31–32, too, ζήτησις must have a broader sense.

stration would beg the question. What could such a demonstration look like? One might think that it could take the form displayed by the following example:

> Human beings are rational animals,
> rational animals are human beings,
> therefore, human beings are human beings.

Nevertheless, it is clear on the face of it that such a syllogism would make very little clear.[26] For this reason, I am not imputing to Aristotle the view that a metaphysician or scientist would actually express the explanation of the existence of a substantial kind by means of a demonstrative syllogism that has as a middle term the essence of that kind. So what is accomplished in determining that the middle term of the existence of a substantial kind S is S itself?

Such an explanation makes clear that the question of whether there is such a thing as S (which *APo.* 2.2.89b37–38 and 90a6–7 identify as the question about whether there is some middle term expressing the cause of S) is to be answered in the affirmative. There is some basic feature of the world responsible for there being a certain substantial kind. It happens to be the substantial kind itself.

What does it mean to say that the final cause of a substantial kind is the kind itself? Here it is helpful to follow Aristotle's lead in first considering products of human art. The final cause of a house is to provide shelter; that of a bed is to provide a support for sleeping. It is clear that it is likewise the case that the final cause of there being any houses at all is to provide shelter; the final cause of there being any beds at all is to provide support for sleeping. Since such things are not by nature, the final cause will be the goal of an entity external to items of the kinds in question. The central insight of Aristotle's natural philosophy is that the final cause of particular natural substances is the formal cause of that substance, which is identical to the full actualization of the essence of that kind to which the substance belongs (*Phys.* 2.1.193b12–18). (In

26. The point can be put otherwise. At *APo.* 2.16.9817 Aristotle claims that "the cause is prior to that of which it is the cause." If a kind is its own cause, it will be both prior and posterior to itself. But I take Aristotle's requirement that the cause be prior to the effect to ensure that the identification of cause renders intelligible the existence of the effect. Substantial kinds are a special case for which the existence is rendered intelligible simply by having the mind focus on the essence of that kind as such.

this case the goal of such an actualization is internal to the substance as such.) We can infer that, from a scientific standpoint, the final cause of a natural substantial kind would likewise be the essence of that kind (which is not to say that, from a metaphysical standpoint, we are entitled to consider kinds the sorts of things that can have a goal).

This discussion is presented in support of the thesis that Aristotle's phrase "that of which there is no other cause" at *Posterior Analytics* 2.9.93b21 refers to those kinds whose members are not considered by the science studying them as inherent in anything more basic. It can be objected that there is no mention of final causation in the *Posterior Analytics* until 2.11; final causation does not seem to play a major role in Aristotle's early account of scientific explanation. But this does not pose a major problem. As we have noted, *Metaphysics* 7.17 is best read as an elaboration on Aristotle's earlier views. The problems with which that chapter deals derive from the application of Aristotle's theory of scientific explanation to metaphysical issues. We can infer that Aristotle had always held that the cause of there being a substantial kind was that kind itself; only in the light of his mature metaphysics is that kind to be identified with its own final cause.

D. Definition and Final Cause in Aristotle's Later Metaphysics

We have seen, according to Ross, the view of *Metaphysics* 7.17 that the essence of a substantial kind is its final cause is in conflict with Aristotle's usual account of the essences of substantial kinds: the genus of that kind conjoined with the differentia. Were this the case, doubt could be cast on the validity of my reading into the *Posterior Analytics* the teaching that the cause of the existence of a kind is the essence of that very kind. But is this the case?

We must first realize that for Aristotle the only real substantial kinds are either immaterial substantial kinds or natural substantial kinds (*Metaph.* 12.1.1069a30–b2). Aristotle never discusses the question of the nature of the *definition* of immaterial substances. (Severe epistemological problems in regard to such substances are at least in part to blame.) For that reason, let us attend to the nature of the definition of composite substances alone.

Although Aristotle does not discuss the metaphysics of change within the *Organon*, according to Aristotle's later metaphysical thought, the process of taking on a certain form, having a certain potentiality actualized, is for the sake of the very form that is being actualized (*Phys.* 2.1.193b13–18). Aristotle clearly indicates that this is so for *natural* substances, that is, those having within themselves as such a certain principle of movement or change. The nature of a thing, its principle of movement or change in the thing as such, is first and foremost its form, which is the actuality for which it strives (*Phys.* 2.1.193b7–18; *Metaph.* 9.8.1050a4–23; *PA* 1.1.641b23–30).[27] This form will be the essence, expressed in the definition of that of which it is the form (*Phys.* 2.1.193a30–31, cf. *Metaph.* 7.7.1032b14).

So we should not be surprised to read in *Metaphysics* 7.17 that the essence of a natural substance should be identified with its final cause. But it is still unclear how this squares with Aristotle's usual understanding of an essence of a kind as the conjunction of the genus and differentia of that kind. We can understand how a thing's purpose is to *do* certain things. But how can it be a thing's purpose to *belong* to a certain genus and have predicated of it certain features that distinguish a thing of its kind from others belonging to that genus?

Here it is necessary to appeal to the distinction between first and second actuality, familiar from *De Anima* 2.1.412a22–27. The first actuality is the actualization of a potentiality to be *able* to do a certain thing. An example of a first actuality is ἐπιστήμη, which, when attained, actualizes the intellect's ability *to be able to* bring before the mind certain truths. For example, one with geometrical knowledge can bring before the mind geometrical truths whenever he or she wishes but need not be actually doing this in order to possess that knowledge at a given time. The contemplation (θεωρία) of geometrical truths is the second actuality; the geometrical knowledge enabling one to engage in geometrical θεωρία when one pleases is the first actuality. The soul of a living being is its form, which Aristotle identifies with its first actuality (*De An.* 2.1.412a22–27). Aristotle's meaning is that, in order to be alive, the body, or matter, of a composite substance must have within it a principle of organization so that it is *able* to of itself engage in those functions in which a living being of that kind can by nature engage. For example, the soul of a human being is the human form that allows the human being to eat, grow, reproduce, move about, desire, imagine, and think.

27. This is why Aristotle takes the final cause of a human being to be the full actualization of that human form. Thus, at *EN* 1.7.1098a16–18 the final cause of all human activities is explicitly identified with happiness (εὐδαιμονία), which is in turn identified with "a life of activity in accordance with virtue."

One need not be actually doing these things in order to be alive, that is, in order to have a human soul, but one must be *able* to do these things when fitting.

If this is the case, we should expect that natural forms, among which are the souls of living beings, will be defined on the basis of the activities that the possession of these forms makes possible. We find that this is indeed the case. In Aristotelian chemistry the basic natural forms are defined in terms of the *motions* things having those forms engage in as well as the *effects* things having those basic forms have on other things. In *De Anima* 2.2–3 Aristotle classifies all living things in a hierarchical manner on the basis of what those living beings can do. The genus of a natural kind is determined on the basis of the natural movements or activities that can be accomplished by a thing of that kind, considered as such. We can see how an essence, considered from one perspective as the *classification* of a kind on the basis of the family the kind belongs to and what sets off that kind from other kinds belonging to that family, can from another perspective be considered the final cause, the conjunction of the first actualities that members of that kind naturally strive to exercise.[28]

We therefore see that Aristotle's view that the essence of the being of a species is the final cause of beings that belong to that species is consonant with his more usual account of the essence of a species as its genus conjoined with its differentia. The assertion of *Metaphysics* 7.17 that the cause of there being a species of a composite substance is the essence of

28. Of course, in the case of living substances only rather inclusive genera (e.g., the genera "plant" and "animal"), as well as the human species (the only rational animal), will be defined on the basis of what those activities are that things of that kind naturally strive to accomplish. But sometimes the species will be distinguished on the basis of unique bodily characteristics (such as size and color of bodily parts) that are incidental to the way in which beings of that kind engage in their life functions. To the extent that a differentia is a determination of characteristics of the bodily parts that are incidental to the way in which those parts function, there may indeed be a problem reconciling the notion of the essence of a kind as its final cause, as expressed in *Metaph.* 7.17, with the notion of essence as determined through genus and differentia. In respect to this two things can be said. First, at *Metaph.* 7.12 Aristotle indicates that a differentia will be a further determination of the genus, so that the differentia, properly understood, will be understood as a differentia of a particular genus. This suggests that, at least within *Metaph.*, Aristotle conceives of a differentia as a further determination of the activities and motions of which that kind is capable. Second, the fact that the many differentiae are not understood in terms of function should not obscure the fact that the most basic genera, as well as some differentiae, are so understood.

that species, which is its final cause, is not inconsistent with anything in the *Posterior Analytics*.

Our consideration of *Posterior Analytics* 2.1–2 has been supplemented by attention to *Metaphysics* 7.17, in order to help determine what Aristotle means when he identifies the scientific inquiry into the existence of a substantial kind with the inquiry into whether there is some middle term expressing the cause of the existence of that kind. We have concluded that Aristotle somehow conceives of the essence of a substantial kind as its own causal middle term. Though this relation cannot be expressed in the form of a demonstration conforming to the requirements for scientific explanation laid down in *Posterior Analytics* 1, the recognition of a substantial essence as its own middle term allows the scientist to answer the question of existence affirmatively: there is *some* feature of the world that is responsible for the fact that there are things of that kind.

In this way the substantial kind is distinguished from what does not exist. The claim that some kinds are their own cause sets these kinds apart from another class as well: those kinds whose cause is expressed by a middle term that is not the same as themselves. These will be the referents of the minor terms of the sort of demonstrations discussed in detail in *Posterior Analytics* 1.

There are three ways in which I shall support this account of the form of the positive resolution of the inquiry into a substantial kind. The first is to show on independent grounds that the theory of science developed in *Posterior Analytics* rests on a distinction between kinds whose existence is to be demonstrated by means of identifying a middle term other than those kinds and those whose existence is not so demonstrated.

I have thus far argued that Aristotle is theoretically committed to the position that certain kinds, which the sciences do not consider as inherent in any more basic subject, must be considered by those sciences as the explanatory ground of their own being. I have not shown how this view squares with Aristotle's explicit assertions concerning the linguistic formulation of demonstrative premises and the demonstrations on which they are built. Showing this is the second way in which I shall provide further support of my position. The task of chapter 3, in which I present and defend an account of the role of assumptions of εἶναι within Aristotle's theory of explanation, is to support my account in these first two ways.

My account would be given further support by showing how it sheds light on aspects of Aristotle's theory that would otherwise be unintelligible. In chapter 5 I provide this third support of my position. I first show how *Posterior Analytics* 2.8 and 2.9 cannot be adequately understood along the lines of the prevailing interpretation. I then argue that the beings identified as having "no different cause" in 2.8 and 2.9 are those kinds that are not taken by the sciences studying them to be inherent in any more basic subjects, those that are considered as being in some way the ground of their own being.

3

Epistemic Substances and Assumptions of Existence

The interpretation of *Posterior Analytics* 2.8–9 to which I am leading rests on interpreting the phrase "that which has no other cause" as referring to those kinds that are not considered by the sciences studying them as inherent in any more basic subject. In chapter 2 I showed how *Posterior Analytics* 2.1–2 and *Metaphysics* 7.17 support the thesis that for Aristotle a kind that is not considered as inherent in a more basic kind would, in some sense, be its own cause. In the present chapter I support this interpretation by attending to a number of distinctions made in *Posterior Analytics* 1. I first present an interpretation of those passages in which Aristotle distinguishes between two different kinds of first principles: assumptions of the εἶναι (being) of something (hypotheses) and assumptions of the essence of something (definitions) (1.1.71a11–17; 2.72a14–24; 10.76a31–36, 76b3–16). I then show how and why in 1.10 (76b6–11, 13, 15–16) Aristotle asserts that existence must be assumed in the case of those kinds that are not taken to inhere in more basic kinds, while those that are understood as inherent in other kinds must have their existence demonstrated. In order to make my case I defend my (traditional) account of the passages concerned with assumptions of εἶναι against recent challenges by Gomez-Lobo and Hintikka. After showing why the former kinds are considered by the sciences studying them to be substantial (though, from a metaphysical point of view, they need not be), I suggest that they be called *epistemically substantial*. I argue that the kinds that are demonstrated to belong to epistemic substances are the καθ' αὑτὰ συμβεβηκότα. I present accounts of the nature of the linguistic formulation and of the logical role of the assumptions of the existence of epistemic substances.

The major passages at issue in this chapter are written in a style even more terse and crabbed than most of the rest of the *Posterior Analytics*. Accordingly, each has been interpreted a number of different ways. To support my case, each passage needs to be examined in turn. Taken in isolation, each passage admits of a number of other possible interpretations, some of which, in some respects, may seem to better fit the particular text in question. But I shall show that, taken together, my interpretations grant to Aristotle a single account of assumptions of existence in scientific explanation that is both internally cogent and grounded in the text.

A. Assumptions of Being in *Posterior Analytics* 1.2

Aristotle's fullest account of the nature of assumptions of εἶναι, at *Posterior Analytics* 1.2.72a14–24, is as follows:

> Among unmediated syllogistic principles, I call posit (θέσις) that principle which is not demonstrable and need not be subscribed to by one who is to learn something (μηδ' ἀνάγκη ἔχειν τὸν μαθησόμενόν τι). But a principle that must be subscribed to by anyone who is to learn something is an axiom. For there are some such things, and we are especially wont to call such principles by that name. Among posits, the principle that takes one or another side of a contradiction (I mean this sort of thing: for something to be or for it not to be) is a hypothesis, but the principle without this is a definition. For the definition is a posit. For the arithmetician posits the monad as being what is indivisible in respect to quantity, but that is not a hypothesis. For what a monad is (τὸ . . . τί ἐστι μονάς) and the being of a monad (τὸ εἶναι μονάδα) are not the same.

Aristotle begins by characterizing a posit as a first principle that is not an axiom (72a14–16). That is to say that it is a proposition that must be assumed in order to learn some things, but not everything. Since, as Aristotle stresses in 1.7, each science must rest on only its own set of first principles (in addition to the axioms; cf. 1.7.75a41–42, 75b2–3) and the subject of every posit will fall under a genus considered by a single science (1.7.75a42–75b12), every θέσις will be a first principle of only one science, that concerning the genus of the subject.

Aristotle subdivides θέσεις into two classes: hypotheses and definitions. Aristotle distinguishes hypotheses and definitions on the grounds that only the first kind of principle "takes one or another side of a contradiction." He then presents an instance of what he means: τὸ εἶναι τι or τὸ μὴ εἶναι τι. This is introduced by the clause οἷον λέγω, which, although literally translated as "I mean, for example," can also have the sense of "id est [i.e.]" or "that is to say."[1] We may therefore take τὸ εἶναι τι or τὸ μὴ εἶναι τι as providing a description of the sort of principle Aristotle has in mind, not as merely providing an example of what has already been definitively described. This is supported by the fact that Aristotle presents an example of a hypothesis (τὸ εἶναι μονάδα) of this form; there was no need to provide another example.

The phrase τὸ εἶναί τι can have three senses. First, τί acts as a variable denoting the subject of the hypothesis and εἶναι has its existential sense. A hypothesis would state that something exists or does not exist.[2]

Second, the indefinite pronoun τί (something) could be acting as a variable that can be filled by a suitable complement of the predicate εἶναι.[3] Aristotle would then be saying that a hypothesis asserts that a subject is an x or is not an x. A hypothesis would then be characterized as a positive or negative predication. To maintain a parallel in the syntax of τὸ εἶναί τι and τὸ εἶναι μονάδα, the latter, which is Aristotle's example of an arithmetical hypothesis, must be translated as "to be a monad." Aristotle would then be saying that the arithmetician hypothesizes about some entity that it is a monad.

The εἶναι is taken to be predicative in the third possible interpreta-

1. On this as a possible translation, see LSJ s.v. οἷος V.2.e. R. McKirahan (*Principles and Proofs: Aristotle's Theory of Demonstrative Science* [Princeton: Princeton University Press, 1992], 42) presents a full discussion of the options the translator faces and a defense of translating the phrase as "i.e., I mean." Others who take the phrase in this way are H. Tredennick (trans. *Aristotle: "Posterior Analytics" and "Topics,"* Loeb Classical Library [Cambridge, Mass.: Harvard University Press, 1960], 33) and W.D. Ross (*Aristotle's "Prior and Posterior Analytics"* [Oxford: Clarendon Press, 1949], 508).

2. This is the line of interpretation followed by Tredennick (*Aristotle*, 33) and McKirahan (*Principles and Proofs*, 41–42). A fourth possible grammatical reading of τὸ εἶναί τι ἢ τὸ μὴ εἶναί τι is "for x to be something or for x not to be something," where τί is not a variable for some predicate complement but means "something at all" as opposed to "nothing." [See B. Landor ("Definitions and Hypotheses in *Posterior Analytics* 72a19–25 and 76b35–72a4," *Phronesis* 26 [1981]: 312)], in which other examples of εἶναί τι being used in this way are presented.) Though the grammatical analysis differs, this analysis gives the passage the same philosophical sense as does the third reading.

3. Examples of this locution within the *APo.* are 1.23.84b30–31 and 1.25.86b8–9. It is true that in these passages there is no article modifying the infinitive, but manuscript C lacks the second τὸ in the passage at hand. See H.W. Smyth, *Greek Grammar*, rev. G. Messing (Cambridge, Mass.: Harvard University Press, 1956), 937a: "An indefinite subject of the infinitive (τινά, ἀνθρώπους) is usually omitted." I am grateful to Thomas Upton for help on this point.

tion as well. But here τί is taken as a variable that fills the place of the subject of some predication, the predicate complement of εἶναι left unstated. Here too Aristotle is taken to be identifying positive or negative predications as hypotheses. But, to maintain a syntactic parallel, the example of the arithmetician's hypothesis must be given a different interpretation: the arithmetician hypothesizes some predication of which the subject is "monad." (Such would be a case of what Aristotle calls εἶναι ἐπὶ μέρους at 2.2.89b39–90a4.)

There is no sure way to determine on the basis of this passage alone what the first principle that Aristotle calls a hypothesis is and what role he takes it to play. So we must turn to other Aristotelian passages concerning the assumption of a thing's εἶναι. We need to find an account that allows a cogent reading of all of these passages considered together and also sheds light on related texts.

B. Assumptions of Being in *Posterior Analytics* 1.1 and 1.10

In *Posterior Analytics* 1.1.71a11–17, Aristotle distinguishes two kinds of prior knowledge that make teaching, learning, and deductive arguments possible. In regard to a given subject, sometimes one must already know ὅτι ἔστι, sometimes "what is that which is said" (τί τὸ λεγόμενόν ἐστι), and sometimes both of these. As an example of a principle for which we must assume ὅτι ἔστι, Aristotle presents the principle of the excluded middle. It is obvious that Aristotle's usage of ὅτι ἔστι here is not consistent with that of 2.1; he is here using ἔστι in its veridical sense, according to which ὅτι ἔστι can be rendered "it is the case." Aristotle's identification of the second kind of object of preexistent knowledge, "what that which is said is," is ambiguous. On the one hand, τὸ λεγόμενον can mean "that which is uttered"; Aristotle would then be demanding that there be a grasp of the meaning of the terms employed in the expository discourse. On the other hand, τὸ λεγόμενον can refer to what is posited as the subject of a statement or discourse,[4] in which case τί τὸ λεγόμενόν ἐστι would be the τί ἐστι, or essence, given in the definition of the subject being considered. This would, among other things, fix the reference of terms employed in the scientific discourse.[5] Aristotle presents an example: while "triangle" is an exam-

4. See LSJ s.v., λέγω III 15.

5. We shall see that Aristotle envisages both the fixing of a term's meaning and the identification of the essence of what a term denotes as definitional principles of the sciences.

ple of that for which we assume only τί τὸ λεγόμενόν ἐστι, we must assume both ὅτι ἔστι and τί σημαίνει (what it means) for "monad."

The most straightforward way of reconciling the present passage with 1.2.72a14–24 is to take the assumption of ὅτι ἔστι discussed here as being the same as the assumption of τὸ εἶναι τι in 1.2. We have seen that the latter can be understood in three ways: as denoting that some given entity exists, as denoting that some subject has this given entity predicated of it, or as denoting that some given entity has some other being predicated of it. The present passage also admits of all three interpretations. On the first interpretation Aristotle says that we assume that there are monads but not that there are triangles. On the second interpretation Aristotle says that we assume of something that it is a monad, but we do not assume that something is a triangle. On the third interpretation Aristotle says that we assume that a monad is something or other but not that a triangle is something or other. At this stage of our analysis the only option that we can securely dismiss is that of taking εἰ ἔστι throughout this passage in its veridical sense (to be true), for a monad and a triangle are not subjects that admit a truth value.[6]

More light is shed on this problem in 1.10, in which Aristotle tells us how it is that the geometrician secures the εἶναι of a triangle. The relevant texts are as follows.

> I call the principles in each genus those whose "that it is" (ὅτι ἔστι) cannot be demonstrated. What the primitives and those that are based on them (τὰ ἐκ τούτων) signify is assumed; that they are must be assumed for the principles, but this is demonstrated for the others. For example, what a monad is, or what "straight" or "triangle" is, must be assumed, as must be the being of the monad and magnitude, but the being of the others must be demonstrated (ἀνάγκη . . . εἶναι τὴν μονάδα λαβεῖν καὶ μέγεθος, τὰ δ' ἕτερα δεικνύναι). (1.10.76a31–36)

6. At 1.1.71a16 ἔστιν cannot have its veridical sense, since its subject is neither a proposition nor a fact but, rather, a single term, substance, or attribute. Further, ἔστι at 1.1.71a14 can have neither a predicative nor an existential sense, as its subject must be construed as a proposition or fact. Since εἶναι is evidently intended to have the same sense in both passages, there is no way to avoid the conclusion that Aristotle has not here clearly distinguished the two senses of εἶναι when he claims that for "monad" *both* kinds of principle just described must be assumed (1.1.71a15–16). Aristotle is not always guilty of this confusion; in *Metaph.* 5.7.1017a31–35 the veridical sense of εἶναι is clearly distinguished from that of the copula.

> Also proper to a science are those things that are assumed to be, whose in-itself predicates (τὰ ὑπάρχοντα καθ' αὑτά) are the concern of the science. For example, arithmetic is concerned with monads, and geometry is concerned with points and lines. For they assume that these are, and that they are *this* (ταῦτα γὰρ λαμβάνουσι τὸ εἶναι καὶ τοδὶ εἶναι). They assume the meaning (τί σημαίνει) of each of the in-itself attributes (πάθη καθ' αὑτά). For example, arithmetic assumes the meaning of "odd" or "even" or "quadrangle" or "cube"; and geometry assumes the meaning of "irrational" or "deflection" or "verging," but that they are (ὅτι δ' ἔστι) they prove through common premises and from what has been demonstrated. Astronomy, too, proceeds in this way. For every demonstrative science is concerned with three kinds of things: those of which it posits the being (those make up the genus of which the in-itself attributes [τὰ καθ' αὑτὰ παθήματα] are matters for study), the so-called common axioms, which first principles are the basis of demonstration, and third, the attributes (τὰ πάθη), of which it assumes what each means. (1.10.76b3–16)

As we shall see, these texts are conclusive evidence against the third way of interpreting assumptions of being of a kind (ὅτι ἔστι, in 1.1, and τὸ εἶναι τι, in 1.2) of those previously proposed: that some kind has some attribute predicated of it. It is not as clear whether these passages are incompatible with either of the other two ways of understanding assumptions of the being of a kind (as an assumption that some given kind exists or as an assumption that some subject has that kind predicated of it). Accordingly, I shall first present an account of these texts that does not decide on this important issue. Following this, I shall take up Gomez-Lobo's defense of the second interpretation and show its defects. I shall then present my account, which follows the first interpretation, and argue against the competing view of Hintikka.

In 1.10 Aristotle first classifies principles (ἀρχαί); he then classifies the things that are the concern of the sciences. What are these classifications, and how are they related?

The term ἀρχαί is ambiguous. On the one hand, ἀρχή is used to refer to a first principle of science, one of the indemonstrable propositions grounding all demonstrations (cf. 1.2.72a38, b2; 1.12.77b5; 1.32.88b21, 27; 2.3.90b24–27). On the other hand, ἀρχή is often used outside of the *Organon* to refer to a simple entity or event responsible for what follows,[7] and in *Posterior Analytics* 2.19, in which Aristotle's explicit concern is with the acquisition of the ἀρχαί, the discussion is focused on the way in which we grasp simple concepts, leading the reader to identify these concepts with the ἀρχαί under discussion.[8]

If we take ἀρχή in the present passage to have its usual sense in the *Posterior Analytics*, an unmediated proposition underlying demonstration,[9] then the first part of the passage admits what is on first consideration a natural interpretation. If we take εἶναι at 76a31–34 to have its veridical sense, Aristotle is distinguishing principles that are assumed to be true in the exposition of a science from those demonstrated to be true.[10] But there are two difficulties with this interpretation. First, it would not be clear what Aristotle would be asserting in his claim that we must assume what both kinds of propositions signify. Is he merely claiming that the exposition of a science presupposes a certain linguistic competence?[11] Second, it is natural to take the distinction between ἀρχαί, which are assumed to "be," and what is based on them, which are demonstrated to be, in such a way as to correspond to the distinction presented as an example in what immediately follows, between subjects for which we assume both the τί ἐστι and the existence and the subjects that are demonstrated to be. But on this interpretation the distinction is only drawn for the sake of the kinds of propositions it brings to light: on the one hand, indemonstrable ones, such as assertions of the τί ἐστι and of the existence of the first kind of subject, and, on the other,

---

7. Thus, at *Metaph.* 5.1.1013a18–20 Aristotle indicates that ἀρχαί, in every sense of the term, are origins of a thing's being, becoming, or becoming known. At 1013a14–17 "hypotheses" (here, having simply the sense of "premises" [on this see n. 25]) are said to be an *example* of ἀρχαί, in the sense of being the starting points of knowledge; the possibility of there being other starting points of knowledge is left open.

8. See J. Barnes, *Aristotle's "Posterior Analytics," Translated with Notes* (Oxford: Clarendon Press, 1975), 249, 259–60.

9. This is the sense in which Aquinas (*In Aristotelis Libros Peri Hermineias et Posteriorum Analyticorum Expositio*, ed. R.M. Spiazzi [Turin, It.: Marietti Editori, 1955], L.I,l.xviii,152) and Ross (*Aristotle's "Prior and Posterior Analytics,"* 538) take this occurrence of ἀρχή.

10. Cf. *Phys.* 2.3.194b23–26 and 195a18–19 (which are the same as *Metaph.* 5.2.1013a24–26 and b20–21), in which εἶναι ἐκ (to be from, to be based on) is employed to express the relation between a deduction's conclusion and its premises, and this is identified as an instance of the relation between a thing and its material cause.

11. Aquinas (*In Aristotelis*, L.I,l.xviii,152) takes Aristotle to be making the claim that, since definitions give the essence of a thing, they presuppose a grasp of what essence and substance are, which grasp is provided by metaphysics. This is unlikely, as there is no evidence in *APo.* that Aristotle envisages all sciences to be in any way subordinate to some single science.

demonstrable ones, such as those asserting the existence of the second kind of subject. This destroys the evident parallelism of the passage and saddles Aristotle with the view that all assertions of the τί ἐστι are indemonstrable, a view that (as we shall see) Aristotle in *Posterior Analytics* 2.8–10 shows to be overly simplistic.

It is best to take the ἀρχαί as constituting a kind of subject, whose being is to be assumed by the science studying them.[12] What does Aristotle mean by calling those subjects whose being must be demonstrated "those based on them" (τὰ ἐκ τούτων)? It has been suggested that these terms are what modern logicians call derived terms, which are defined on the basis of primitive terms (the ἀρχαί).[13] This is supported by *Metaphysics* 5.24.1023a35–b2: "Those things are based on something that are like the form based on the part, for example, human is based on 'two-footed' and syllable is from 'letter'"; these are said to be examples of how "the form comes from the matter of the form" (τὸ εἶδος ἐκ τῆς τοῦ εἴδους ὕλης). But, as Barnes points out,[14] while what modern logicians call primitive terms are undefinable, Aristotle claims that for both of these kinds of subjects there is a signification (τί σημαίνει) that is expressed in an assumption.

We must take "to be based on" in the sense of "to be deduced from."[15] The distinction between the principles and what is based on

12. Here I follow McKirahan (*Principles and Proofs*, 40).

13. See H. Scholz, "The Ancient Axiomatic Theory," trans. J. Barnes, in *Articles on Aristotle*, vol. 1: *Science*, ed. J. Barnes, M. Schofield, and R. Sorabji (London: Duckworth, 1975), 52, 60.

14. Barnes, *Aristotle's "Posterior Analytics,"* 133–34.

15. McKirahan (*Principles and Proofs*, 40–41) translates τὰ ἐκ τούτων as "derivative things," surely an overtranslation, especially since the distinction between primitive subjects and derivative subjects, never explicitly drawn by Aristotle, is a cornerstone of McKirahan's own interpretation of Aristotelian first principles (36–49). I see no way in which a sharp distinction can be made between what McKirahan calls "derivative subjects" and "attributes." McKirahan recognizes that, like derivative subjects, attributes can be said to exist: "for an attribute to exist is for it to belong to a subject" (196). Why cannot a derivative subject be considered the attribute of standing in a certain relationship with other subjects? If there is no reason why this cannot be so, then a derivative subject is a kind of attribute. To this McKirahan might reply that derivative subjects can themselves serve as subjects for attributes, while attributes cannot. But any attribute A′ of an attribute A can be considered merely as a *determination* of A. For example, let us take a triangle as an attribute of a line, namely, the attribute of standing in a certain geometrical relation with two other lines. Then to predicate "isosceles" of "triangle" is simply to describe with greater specificity the relation that the line that is described as subject stands to the other two lines.

them is the distinction between those things whose being is assumed and those whose being is deduced on the basis of assumptions of being. This allows us to make sense of the example to which Aristotle points. The being of a magnitude must be assumed; the being of a triangle must be demonstrated on the basis of the assumption of the being of a magnitude. (On the first of our competing interpretations, this would mean that the geometer assumes the existence of magnitudes and proves that there are such things as triangles. On the second competing interpretation, it would mean that through an assumption a geometer takes some entity [or entities] to be magnitude[s], while it is proven that some entity is a triangle.)

This interpretation is confirmed by what follows, in which Aristotle distinguishes two kinds of subjects that are "proper" to the sciences, those that are assumed to exist and whose καθ' αὑτά attributes are studied, and the πάθη καθ' αὑτά (in-itself attributes) of those subjects, whose existence is demonstrated (1.10.76b3–11). Aristotle says that of the first kind of subject, examples of which are "point" and "line," one assumes "the being and the being *this*" (τὸ εἶναι καὶ τοδὶ εἶναι) (1.10.76b6), that is, the being and the definition of the subject. Only the meaning of the term is said to be assumed of the second kind of subject, of which Aristotle presents "even," "odd," "quadrangular," and "cube" as examples. This distinction corresponds to that between the principles and the things based on them drawn earlier in the chapter; the examples make clear that τὰ πάθη καθ' αὑτά are to be identified with τὰ ὑπάρχοντα καθ' αὑτά of the proper subjects considered by the sciences. The context also makes clear that these are the καθ' αὑτὰ συμβεβηκότα, those attributes of a subject whose inherence in a subject can be an object of scientific understanding but are not implicitly definitional of that subject. It is of such entities that the being is demonstrated, by means of a demonstration that the entity is predicated of one of those things whose being is assumed. This is conclusive evidence against the first of the possible interpretations of assumptions of being that I have sketched. The attributes of a kind studied by a science are *demonstrated* to be predicated of that kind; these predications are not to be assumed. This should not be surprising, for what could the object of demonstration be if not such predications?

Aristotle next identifies the subject of a science, of which the being must be hypothesized, as the genus, of which the παθήματα καθ' αὑτά are demonstrated (1.10.76b11–16). Hintikka has taken *genus* here to refer to the highest genus considered by a science.[16] I will shortly examine the details of Hintikka's interpretation. Here it should suffice to make two remarks. First, Aristotle does not take *genus* and *species* to be mutually exclusive terms. A genus, like a species, is a kind, for which there are certain proper principles. What is to constitute the genus is to some degree dependent on the purposes and scope of one's inquiry.[17] Second, even if we do take Aristotle to be maintaining that, for each of the sciences there is a single genus that is to be hypothesized, that cannot be interpreted to the effect that there will be only one kind whose being will be assumed among the first principles of that science. Aristotle's wording indicates otherwise: "For every demonstrative science concerns three kinds of things, first, all of those subjects that are posited as existing (these are the genus, of which the καθ' αὑτὰ παθήματα are studied . . .)" (1.10.76b11–13). Here a multiplicity of subject terms taken together are said to constitute the genus. This suggests that the hypothesis of the being for some genus G involve hypotheses of the being of species falling under G.[18] To hypothesize the genus of fish is to hypothesize the being of guppies, flounder, and all of the species of fish of which the ichthyologist is aware.

Aristotle's examples in the passage we are presently considering can be understood as follows. The genus of arithmetic is quantity, but being is hypothesized in the case of a *monad;* this hypothesis is a first principle of arithmetic (*APo.* 1.1.71a15–16, 10.76b4–5). The genus of geometry is magnitude, but it is lines and points of which the being is assumed through as first principle (1.10.76b5). Are we to understand the monad as a species of quantity and lines and points as species of magnitude? Here, as elsewhere when discussing mathematical demonstrations,

16. See J. Hintikka, "On the Ingredients of an Aristotelian Science," *Nous* 6 (1972): 62–66.

17. See P. Pellegrin, *Aristotle's Classification of Animals,* trans. A. Preus (Berkeley: University of California Press, 1986), 50–112; D. Balme, "Aristotle's Use of Division and Differentiae," in *Philosophical Issues in Aristotle's Biology,* ed. A. Gotthelf and J. Lennox, (Cambridge: Cambridge University Press, 1987), 71–73.

18. This is also the view of I. Mueller ("On the Notion of a Mathematical Starting Point in Plato, Aristotle, and Euclid," in *Science and Philosophy in Classical Greece,* ed. A.C. Bowen, [New York: Garland, 1991], 68). Aristotle's assertion that the major term of a demonstration must be from the genus of the science by virtue of which one demonstrates (1.7.75b11–12) may simply be an example of loose writing; it is slender evidence for the view that the genus of a science includes the demonstrated attributes, as it does on McKirahan's view (*Principles and Proofs,* 51–52, 60–63).

Aristotle does not provide details, but I think that this must be the gist of his thought. It is not unlikely that, at least in the beginning of his career, Aristotle, like Plato, saw the arithmetician to be positing a field of undifferentiated unities and investigating the properties of the various assemblages of these unities.[19] If so, the various numbers will be considered *attributes* of these monads, considered as grouped in various ways. Thus, attributes such as being odd and being even (see 76b8) will be attributes of attributes of monads (in the last analysis, attributes of monads, grouped together in certain ways). Similarly, geometry will posit the being of lines (with perhaps straight, curved, and various lengths differentiating the species of lines); the geometrical figures will be attributes of lines, considered as assembled in various ways. The attributes of such figures (see 76b9) would then in the last analysis be attributes of lines.[20]

According to the first of our two possible ways of interpreting assumptions of being, Aristotle is here saying that the arithmetician assumes the existence of monads and demonstrates the existence of certain numbers and then, on that basis, the existence of attributes of these numbers. Similarly, the geometer assumes the existence of lines and demonstrates the existence of figures (such as triangles) and then, on that basis, the existence of attributes of these figures. On the second possible way of interpreting Aristotle, he is here saying that the arithmetician assumes that certain entities are monads and demonstrates that certain other entities are numbers and then, on that basis, demonstrates that certain other entities are attributes of numbers. Similarly, on this understanding the geometer assumes that certain entities are lines and demonstrates that certain other entities are figures and then, on that basis, demonstrates that other entities are attributes of figures.

---

19. See *Rep.* 525b11–526b4 and the discussion of both this passage and the Aristotelian understanding of number in J. Klein, *Greek Mathematical Thought and the Origin of Algebra*, trans. E. Brann (Cambridge, Mass.: MIT Press, 1968), 23–24, 46–60.

20. In several passages a triangle is indeed taken to be an essential attribute of certain lines, for its being must be demonstrated (1.1.71a14–15; 1.10.76a35, 2.6.92b15–16; cf. Euclid, *Elements*, theorem 1:1). Elsewhere, though, a triangle is considered the *subject* of certain essential attributes (such as having the sum of the interior angles equal to two right angles) (1.4.73b30–32, 1.23.84b3–9); see Ross (*Aristotle's "Prior and Posterior Analytics,"* 505). This is best considered a kind of shorthand (as discussed in n. 15). The attribute of having interior angles equal to two right angles would be properly considered an essential attribute of certain *lines* arranged in a certain way, but it is not always necessary for the geometer to analyze the attribute to this extent.

C. Gomez-Lobo's Challenge to the Existential Reading of τὸ εἶναί τι and εἰ ἔστι

We have seen that, within the first book of the *Posterior Analytics*, Aristotle distinguishes those beings that are the subjects of the sciences from those that are καθ' αὑτὰ συμβεβηκότα of these subjects on the grounds that the first are assumed to be and the latter are demonstrated to be. The sense we are to give to *to be* here is left unresolved. On the first interpretation εἶναι has its existential sense; Aristotle is saying that each science assumes that the kinds that are its subject matter exist and demonstrates the existence of the καθ' αὑτὰ συμβεβηκότα. Such a demonstration would have as its conclusion a proposition in which the attribute in question is predicated of its subject. On the second interpretation εἶναι has its copulative sense. Aristotle is saying that the demonstrative reasoning proper to some science has as one kind of first principle the assumption that something belongs to or has predicated of it a kind that is a subject of that science. On this view, it is to be demonstrated that this thing has the καθ' αὑτὰ συμβεβηκότα attributes of that kind.

The issue is important to the present study primarily because, if it can be ascertained that the first interpretation is correct, the relevant passages would provide further support of my contention that, within the first book of the *Posterior Analytics*, Aristotle distinguishes between two kinds of beings on the basis of the sort of *explanation* proper to them. On this reading Aristotle distinguishes between those beings whose existence is to be rendered intelligible only on the basis of their own essences (for, metaphysically, they have no cause other than their own essences), and those whose existence is to be explained on the basis of other entities. This distinction in turn is a crucial support of my understanding of *Posterior Analytics* 2.8–10. On the other hand, if the second interpretation is correct, Aristotle nowhere explicitly makes the distinction between the two kinds of beings that I attribute to him. Instead, in the passages under consideration he makes a *methodological* point concerning our *identification* of particulars as falling under certain kinds. On this reading Aristotle distinguishes between those predicates for which the *particular* exemplifications can be identified as such without a justificatory account and those predicates whose inherence in given particulars must be proved through a demonstration. Such a distinction has no bearing on the project of *Posterior Analytics* 2.8–10, as I understand it.

This second understanding of the role Aristotle gives to assumptions of being has been put forward by Gomez-Lobo in two important articles: "Aristotle's Hypotheses and the Euclidean Postulates"[21] and "The So-Called Question of Existence in Aristotle, *An. Post.* 2.1–2."[22] I here discuss and argue against this view.

Gomez-Lobo follows the lead of Kahn in detecting a predicative use of εἶναι in many passages in which an existential use has been taken for granted. He interprets Aristotelian hypotheses as statements asserting that a given substrate is a certain kind of thing, where this identification of a thing makes possible a demonstration that certain attributes inhere in the thing. Accordingly, in "Aristotle's Hypotheses" he argues that τὸ εἶναι τι ἢ τὸ μὴ εἶναί τι at *Posterior Analytics* 1.2.72b18–20 is to be interpreted as "is it an x or is it not an x," comparing the hypotheses, which assert one or the other of these, not with those Euclidean postulates in which the possibility of constructing a certain figure is assumed, but with the ἔκθεσις of a Euclidean proof, in which a given line or figure is assumed to be of a certain kind (e.g., Let AB be a line . . .).[23] In "The So-Called Question" Gomez-Lobo extends this analysis to the assertions of εἰ ἔστι discussed in 2.1–2. He here argues that Aristotle is claiming that some matter must be identified as exemplifying a certain kind prior to an inquiry into the essential attributes of that kind.[24] I shall here examine Gomez-Lobo's position by first considering his arguments against interpreting Aristotelian hypotheses as assertions of existence and then

21. A. Gomez-Lobo, "Aristotle's Hypotheses and the Euclidean Postulates," *Review of Metaphysics* 30 (1977).

22. A. Gomez-Lobo, "The So-Called Question of Existence in Aristotle, *An. Post.* 2.1–2," *Review of Metaphysics* 34 (1980).

23. Gomez-Lobo, "Aristotle's Hypotheses," 438.

24. Hence, Gomez-Lobo must claim that the subject of a hypothesis is a particular, not a universal, concept or form. Upton ("Imperishable Being and the Role of Technical Hypotheses in Aristotelian Definition," *Nature and System* 2 [1980]) and Landor ("Definitions," 318 n. 26) argue that this alone should render Gomez-Lobo's interpretation invalid, since Aristotle takes demonstrations to concern imperishable universals; premises of demonstration must therefore be both eternal and universal. But one holding Gomez-Lobo's view could rejoin that a hypothesis is a taking of a chance case, on the basis of which universal truths are made clear (cf. *APo.* 1.10.76b39–77a3). Gomez-Lobo ("Aristotle's Hypotheses," 436–38) himself regards the taking of chance cases in the ἔκθεσις of a Euclidean proof as very much like an Aristotelian hypothesis.

by evaluating the alternative interpretation he gives and his arguments in support of it.

One of Gomez-Lobo's key arguments is that assertions of existence cannot be employed as syllogistic premises, while Aristotle claims that "hypotheses are among the premises," at 1.10.76b36.[25] But here, unlike at 1.2.72a20, "hypothesis" is being used simply to mean "premise."[26]

Gomez-Lobo claims that the interpretation of εἰ ἔστι at 2.1.89b24 as "whether it exists" commits one to an analysis of an εἰ ἔστι statement as "only one term, whether singular or general, and a verb expressing existence." He claims that this analysis is incompatible with *Posterior Analytics* 2.2, according to which an assertion of εἰ ἔστι is to be explained by virtue of some middle term.[27] But, as we have seen in chapter 2, Aristotle recognizes that only some assertions of existence are properly considered as a predication of one term of another and that a demonstration is possible only of this kind of existence claim.

Gomez-Lobo disputes Ross's suggestion that the subjects of assertions of εἰ ἔστι, as considered demonstrable in *Posterior Analytics* 2.2, are restricted to "attributes and events," excluding subject terms. In support of his assertion that the assertions of a thing's εἶναι concern substantial terms too, Gomez-Lobo calls attention to the mentioning at 2.2.90a4–5, 10, and 12–13 of substance or particular substances as examples of things whose being is ascertained through some middle term.[28] To see what Aristotle is up to in these lines, we need to reconsider them in context.[29]

At 2.2.89b38–39 Aristotle claims that knowledge of "whether something is the case or whether it is, either partially or in an absolute sense" (ἢ τὸ ὅτι ἢ εἰ ἔστι, ἢ τὸ ἐπι μέρους ἢ τὸ ἁπλῶς) must precede any inquiry for a middle term making possible either an account of what something is or an explanation of why something is the case. He then proceeds to explain what he means by these two kinds of being: "By

25. Ibid., 432.

26. ὑπόθεσις has the sense of "premise" at *APr.* 1.23.40b25–29, 41a23–27; 1.44.50a16–b4; *Top.* 3.6.119b35–39; *Phys.* 2.3.195a18–19; and *Metaph.* 5.1.1013a18–20, 13.9.1086a15. On this sense of ὑπόθεσις, see H. Bonitz, *Index Aristotelicum*, rev. O. Gigon (Berlin: Groyten, 1961), 796b59–797a15, K. von Fritz, "Die APXAI in der griechischen Mathematik," *Archiv für Begriffsgeschichte* 1 (1955): 37; on its having this sense at 1.10.76b35–36, see ibid., 40–42; also Barnes, *Aristotle's "Posterior Analytics,"* 136–37 and McKirahan, *Principles and Proofs*, 47.

27. Gomez-Lobo, "So-Called Question," 72–73.

28. Ibid., 73–74.

29. On these lines, see also chap. 2, sec. A.

'partially' and 'in an absolute sense,' I mean this. 'Partially': Does the Moon undergo an eclipse or wax? . . . 'In an absolute sense': whether the Moon or night is or not" (2.2.90a1–5). It is clear from the context that "the Moon is" or "night is" are meant to be examples of a use of "to be" to be contrasted with its predicative use. There is no need to see "the Moon is" (ἡ σελήνη ἐστιν) as providing an example of Aristotle's immediate concern: propositions of the form "ἐστι x" that are demonstrable given the appropriate middle term. The case is similar when Aristotle contrasts "being in an absolute sense" with being not in an absolute sense as follows: "I mean by 'to be in an absolute sense' to be the subject, for example, Moon, Earth, Sun, or triangle, but by 'to be something' I mean to be an eclipse, equal, unequal, whether it is in the middle or not" (2.2.90a12–14). Again, Moon, Sun, Earth, and triangle function as grammatical subjects, able to take the predicates "eclipse," "in the middle," "equal," or "unequal."[30] There is no need to see the particular subject terms here presented as examples of those subjects whose existence is demonstrable.

But exactly what is being in an absolute sense? According to the traditional interpretation, the phrase means "existence." Support for this may be found in Aristotle's examples: an assertion of εἶναι mentions only one subject or attribute, while an assertion of εἶναι ἐπὶ μέρους mentions two, the subject and the predicate. In response Gomez-Lobo points out that Aristotle identifies the middle term as the cause of something's being, in both senses of *being*, as follows: τὸ γὰρ αἴτιον τοῦ εἶναι μὴ τοδὶ ἢ τοδὶ ἀλλ' ἁπλῶς τὴν οὐσίαν, ἢ τοῦ μὴ ἁπλῶς ἀλλά τι τῶν καθ' αὑτὸ ἢ κατὰ συμβεβηκός, τὸ μέσον ἐστίν (2.2.90a9–11). Remarking that τὴν οὐσίαν is omitted without textual support by Mure and Barnes (both of whom take εἶναι ἁπλῶς as "to exist"), Gomez-Lobo offers as a translation, "For the cause of (something's) being not this or that but (of its being) *simpliciter* the substance (involved in the inquiry) or of (its being) not *simpliciter*, but (of its being) one of the things which belong to it *per se* or *per accidens*—that is the middle term."[31] If this is the appropriate translation, the distinction between assertions of εἶναι ἁπλῶς and εἶναι ἐπὶ μέρους is to be taken as between two different kinds of predications; in the former case the predicate is a substantial term,

30. As A.P.D. Mourelatos has pointed out to me, Aristotle here presents a series of correlative subjects and predicates.

31. Gomez-Lobo, "So-Called Question," 87.

while in the latter it is a nonsubstantial καθ' αὑτό term or an accidental term.

There are problems with Gomez-Lobo's translation. First, it is not clear what sense Gomez-Lobo gives to ἁπλῶς here and how ἁπλῶς is to be contrasted with ἐπὶ μέρους. The terms themselves point to a distinction between a general and a restrictive view of a thing's being. When restrictively considering a thing's being, we focus on one of its καθ' αὑτό or accidental predicates. But on Gomez-Lobo's reading a nonrestrictive view of a thing's being, too, focuses on a single predicate of that thing, the substantial kind under which it falls. To be sure, this predicate is of the utmost importance to the science considering that subject, but is not such a perspective, which excludes consideration of accidental predicates, just as restrictive as one that focuses on a nonsubstantial καθ' αὑτό attribute of the subject?

Further, even without the excision of τὴν οὐσίαν, another translation is possible. We can take ἁπλῶς as in opposition to τοδὶ ἢ τοδί (thus preserving the expected contrast of restrictive and nonrestrictive kinds of being) and τὴν οὐσίαν as an accusative of respect. Aristotle would be saying "for the cause of its being not this or that but of its being in an absolute sense in respect to its essence . . . is the middle term." All entities, both those with a middle term and those without, have some cause that can be regarded as a middle term.

Gomez-Lobo bolsters his interpretation of εἰ ἔστι as "whether it is an x" by a consideration of 1.1.71a24–27. In this passage Aristotle considers the case of one who has a grasp of a universal predication but is as yet ignorant of some particular thing falling under this predication's universal subject. Gomez-Lobo writes:

> He illustrates his thesis by means of the following question: "if of a given thing he did not in an unqualified sense of the term know εἰ ἔστι, how could he know with qualification that it has angles equal to two right angles?" . . . Aristotle's point is that knowledge of the universal premise (x)(Tx → Rx) . . . does not involve unqualified knowledge of the conclusion Ra . . . unless one knows a further premise without qualification. Since knowing without qualification is contrasted with knowing universally (71a28–29), the missing particular premise is obviously Ta . . . because knowledge

> of the existence of the figure a . . . is insufficient by itself to make Aristotle's example a valid syllogism.[32]

Taking the passage out of context, Gomez-Lobo misses the point of Aristotle's remarks. At 71a17–24 Aristotle makes the point that, although all learning depends on "prior" knowledge, the attaining of this prior knowledge and the act of learning that it makes possible can sometimes be simultaneous. Thus, in Gomez-Lobo's terms, given $(x)(Tx \rightarrow Rx)$, one will learn Ra as soon as one is aware that Ta. There is nothing problematic in the logical structure of this syllogism, and it would be odd if Aristotle were to devote the next several lines to emphasizing the dependence of the conclusion on the major premise. Rather, his concern is with the very possibility of grasping the minor premise without an awareness of every subordinate kind falling under that premise's subject. We should remember that Aristotle would not express this premise as $(x)(Tx \rightarrow Rx)$. For Aristotle the proposition "All T's are R's" is a predication; at issue is its subject. If the subject is *every triangle* (πᾶν τρίγωνον [71a19–20]), then must we not know that every kind of triangle exists in order to know the premise? (Cf. 1.1.71a30–33.) "For how can it be that what he didn't know to exist he knows to have two right angles without restriction (ἁπλῶς)?" This question, whose context is clearly dialectical, is the one cited by Gomez-Lobo in the quoted passage. Aristotle resolves the problem, which he identifies as a version of the *Meno* problem, by denying that a universal premise is grasped without restriction (ἁπλῶς) (71a27–28). The subject term of a universal proposition regarding triangles is not *every triangle* but *triangle* itself. The proposition is grasped *as* holding of every case and thus allows us to know that a certain predicate inheres in anything we know to be a triangle.

We may further note that, according to the results of Kahn's preliminary study regarding the position of εἶναι, if the phrase τὸ εἶναί τι at 1.2.72a20 does mean "to be an F," as Gomez-Lobo takes it, Aristotle would be employing an unusual locution.[33]

---

32. Ibid., 77–78.

33. "Where no subject is expressed, the predicate nearly always precedes the copula" (C. Kahn, *The Verb "Be" in Ancient Greek* [Dordrecht, Neth.: D. Reidel, 1973], 429).

I conclude that Gomez-Lobo does not present arguments that warrant abandoning the traditional interpretation of hypotheses and assumptions of εἶναι as existence claims.

D. Hypotheses Are Assumptions of Existence

The following view alone remains viable. According to this view, in the passages of *Posterior Analytics* 1.1, 1.2, and 10 that we have examined, Aristotle distinguishes two different kinds of being studied by a science on the basis of the sort of existence claim to be made in regard to each. The only beings whose existence is assumed are those that make up the subject matter of the science. These are those kinds that fall under what Aristotle calls the genus studied by the science (1.7.75a42, 10.76b13). The science considers these the substrates of the attributes it studies. The other sort of entity studied by a science is an attribute that is καθ' αὑτὸ συμβεβηκός of a subject of a science (1.7.75b1–2). It must have its existence demonstrated by means of a demonstration whose conclusion is the predication of that entity of its proper substrate.[34] (Of course, many implicitly definitional predications are also in some sense demonstrable—e.g., "Every man has a backbone, everything that has a backbone is an animal; hence every man is an animal"—although this will not be a demonstration τοῦ διότι. The point is that this proposition *need* not be demonstrated. Accordingly, the existence of this predicate is not to be demonstrated in a properly expounded science.)

On this understanding Aristotle's mathematical examples can be understood as follows. The arithmetician assumes the existence of monads and demonstrates the existence of certain attributes of numbers by demonstrating that they are predicated of monads (assembled in certain ways). Likewise, the geometrician assumes the existence of points and lines and demonstrates the existence of figures (such as triangles) and of attributes of figures (as well as the attributes of other assemblages of geometrical primitives).

Two problems arise. First, it is not clear that ancient Greek mathematicians ever presented proofs of this form. Second, it is likewise

34. Cf. M. Ferejohn, "Definition and the Two Stages of Aristotelian Demonstration," *Review of Metaphysics* 36 (1982): 395. As we shall see (chap. 4, sec. E), the aporia of *APo.* 2.7.92b12–18 rests on the blurring of this distinction, which is again drawn in the solution to the aporia in 2.8. For this reason, no conclusions concerning Aristotle's views should rest on the former passage alone.

unclear why any science would require any assumption of existence as a first principle and what role it would play in demonstrations. I deal with the first issue in the present section and the second issue in the next section.

I here discuss Aristotle's examples of geometrical assertions of existence. The traditional view is that Aristotle's account of geometrical demonstration conforms very well to the geometrical proofs we find in Euclid's *Elements*. According to this view, the assumptions of the existence of certain geometrical primitives correspond to those of Euclid's postulates that are assumptions that a certain geometrical entity can be *constructed*. Likewise, a proof that a certain geometrical entity exists corresponds to a Euclidean construction, that is, a proof that one can construct a geometrical entity conforming to a certain description. Aristotle's main geometrical example of the two kinds of assertions of existence (that one assumes the existence of a line and demonstrates the existence of a triangle) is directly illustrated in book 1 of the *Elements* by postulates 1 and 3 ("to draw a straight line from any point to any point" and "to describe a circle with any center and distance"), on the one hand, and, on the other, by proposition 1 ("On a given straight line to construct an equilateral triangle"),[35] which is proved on the basis of these two postulates.[36]

This view has been challenged. For one speaks of the construction of a geometrical figure as a kind of γένεσις, bringing into being some particular that previously was not. (Aristotle's arguments concerning infinity in *Physics* 3 and 6.2 make clear that he holds that a magnitude is actualized only insofar as there is effected either a mental or a physical *cut*. I take it that when the cut is conceptual this kind of γένεσις would be the actualization of a potentiality in intelligible matter.) In Euclid's *Elements* a kind of construction is proved to be possible if it is shown that one can bring about a particular instance of the construction of the kind in question, with no mention of the individuating characteristics of the geometrical entity that has been constructed. Without recourse to what today's logicians call existential generalization, this procedure does not adequately show that *any* entity of the kind in question can be

35. The translations are from T. Heath, *Euclid's "Elements," Translated with Introduction and Commentary* (New York: Dover, 1956), 1:154, 241.

36. See ibid., 146; cf. Heath, *Mathematics in Aristotle* (Oxford: Clarendon Press, 1949), 56; H.D.P. Lee, "Geometrical Method and Aristotle's Account of First Principles," *Classical Quarterly* 29 (1935): 115; and McKirahan, *Principles and Proofs*, 133–37, 144–47.

constructed. But no Greek mathematician or logician seems to have devised a kind of proof that adequately addresses the problem.[37]

On the traditional interpretation (for which I have been arguing), it is not the coming-into-being of certain entities that is assumed by a science but, rather, their existence. This existence would have to be eternal (excluding the possibility of coming-to-be), since in 1.8 Aristotle emphasizes that scientific understanding must be concerned with eternal kinds. Likewise, it is not the coming-into-being (or construction) of derivative entities that is to be proved, but their (eternal) existence. This is one reason why it is maintained that the mathematical assumptions of existence to which Aristotle refers are a very different kind of assumption than a Euclidean postulate of constructibility.[38]

We can safely assume that the author of *Posterior Analytics* 1 would have been sensitive to the deficiency of geometrical proofs that argue from a single case. The fact that mathematics, like any other science, deals with kinds (cf. 1.13.79a7–8) that eternally exist (i.e., that are eternally exemplified) would be enough to arouse Aristotle's suspicion. Given the absence in antiquity of any other kind of geometrical proof determining which entities are such as can exist and which cannot, it is a safe assumption that, in pondering geometrical examples, Aristotle has in mind constructions of the sort proved by Euclid. But, instead of attending to the logical form that these proofs actually have, Aristotle mistakenly takes for granted that in respect to logical form they conform to his account of demonstration. Aristotle interprets the assumption that a certain line can be constructed as the positing of the existence of all such lines. He likewise takes the construction of a certain figure as a demonstration that all such figures do exist, not realizing that at the very least one needs here a principle of existential generalization.

If this suggestion is correct, the ill fit between Aristotle's account of

37. I. Mueller (*Philosophy of Mathematics and Deductive Structure in Euclid's "Elements"* [Cambridge, Mass.: MIT Press, 1981], 11–15) presents a clear account of the problem and explains how Hilbert's axiomatization of Euclidean geometry avoids this problem, since it explicitly employs the strategy of existential generalization. Interestingly, Hilbert follows the Aristotelian pattern of proving the existence of certain geometrical entities on the basis of assumptions of the existence of certain other geometrical entities.

38. W. Kullman ("Die Funktion der mathematischen Beispiele in Aristoteles' *Analytica Posteriora*," in *Aristotle on Science: Proceedings of the Eighth Symposium Aristotelicum*, ed. E. Berti [Padua: Editrice Antenore, 1981], 247–50) draws attention to the discrepancy. The problem would dissolve if McKirahan, *Principles and Proofs*, 128–32) is correct that Aristotle is not committed to the eternal existence of the kinds studied by the sciences; in sec. E, I argue that Aristotle is committed to this.

geometrical hypotheses and the geometrical assumptions actually employed by ancient mathematicians does not tell against interpreting Aristotle's assumptions of being as assumptions of existence.

E. Why Are There Assumptions of Existence?

Why does Aristotle posit the assertion of existence as a kind of first principle? All of the unmediated predications that serve as ultimate premises of explanatory demonstrations can be derived from definitions. It appears that there is no need for any first principles other than definitions and the laws of inference.

In this section I present the following solution to the problem. For Aristotle the truth of any predication entails the existence of its subject. This is implicit within the logic of the *Prior Analytics* and is made explicit in *Categories* 10. Aristotle is aware, however, of nominal definitions of fictional or hypothetical entities, which are true in some sense, insofar as they accurately describe how language is or ought to be employed. Nonetheless, nominal definitions of nonexistents are not to be admitted as premises of demonstration, for in that case the conclusion of the demonstration would be false, having a subject term that does not denote any existing thing. Hence, there must be some device employed within the exposition of a science to distinguish a nominal definition from a definition of a real kind that is otherwise identical in form. In those cases in which the assertion of the subject's existence is to be regarded as a mediated predication, this will be accomplished by a demonstration of that predication; given such a demonstration, one knows that the definition of the kind is a true proposition, reflective of reality. But, in the case of the basic subjects of the sciences, an explicit acknowledgment of the subject's existence is required.[39] This assertion of existence will not function as a premise of demonstration, but it will be a principle, or starting point of demonstration, since only such an assertion makes possible demonstrations concerning that subject.[40]

---

39. On the role of existence claims, I am fundamentally in agreement with McKirahan (*Principles and Proofs*, 124), who writes: "Existence claims and definitions cooperate in the task of introducing subjects into a science. The existence claim simply states that the subject exists; the definition informs us what kind of subject it is. Together they legitimize the subject as a member of the subject genus and locate it in the genus."

40. Though our accounts of Aristotle's classification of demonstrative principles differ in a number of respects, on this point I am in agreement with Mueller. He writes: "My proposal . . . is that we separate the purely logical notion of a principle as an ultimate

The exposition of my solution presupposes some grasp of Aristotle's semantics and the import of his requirement that the subject of every proposition refers. The laws of conversion set forth in *Prior Analytics* 1.2.25a1–26 hold only given that the subject of any universal proposition refers. Aristotle claims that from a universal proposition, such as "all A's are B's," there follows the particular proposition "Some B is an A." But given $\forall x(Ax \rightarrow Bx)$, we cannot infer $\exists x(Bx \,\&\, Ax)$ without the additional premise $\exists x(Ax)$. The Aristotelian premise "All A's are B's" must then be treated as $\forall x(Ax \rightarrow Bx) \,\&\, \exists x(Ax)$ to make the rules of conversion valid.[41]

Lukasiewicz has argued that Aristotle has not erred in positing his laws of conversion on the grounds that empty terms and quantifiers are absent from his logic, which concerns "only universal terms, like 'man' or 'animal.'"[42] He could have added that such universal terms are assumed to have reference in the discourse within which they are employed.

But Aristotle is not unaware that reference is sometimes made to fictional or otherwise nonexistent entities or kinds; how are propositions concerning such entities or kinds to be considered? There are two possibilities here. Either they are meaningless, or they are false. This issue was at the center of much debate in early analytic philosophy (with Fregeans arguing that they are meaningless and Russellians that they are false), for the issue has great bearing on that of the nature of reference itself. Where does Aristotle stand on this question?

If "All A's are B's" does indeed have the same truth conditions as $\forall x(Ax \rightarrow Bx) \,\&\, \exists x(Ax)$, then, given $\sim\exists x(Ax)$, "All A's are B's" will be false. This is confirmed by *Categories* 10, which is concerned with the kinds of opposition. Here Aristotle writes:

---

premise of proof from what might be called the analytic notion of a starting point, analytic because it is based on an analysis of mathematical practice. The two notions coincide insofar as one kind of starting point is an ultimate premise and insofar as neither starting points nor ultimate premises are provable, but they do not coincide completely since not all starting points are ultimate premises" (I. Mueller, "On the Notion of a Mathematical Starting Point," 81).

41. That is not to say that $\forall x(Ax \rightarrow Bx) \,\&\, \exists x(Ax)$ is a *translation* of "All A's are B's." For the former expression expresses a truth about all entities, not just about A's. Henry Mendell first indicated this to me; I later saw much the same point made in Ferejohn, *Origins*, 43–44).

42. J. Lukasiewicz, *Aristotle's Syllogistic from the Standpoint of Modern Formal Logic*, 2d ed. (Oxford: Clarendon Press, 1957), 130–32.

> It is clear that all of the pairs that are opposed as affirmation and negation are opposed in none of the aforesaid ways. For in these cases alone is it always necessary that one of them be true and the other false . . . This would especially seem to be the case when contraries are said in composition with other terms. For "Socrates is healthy" is contrary to "Socrates is ill." But in these cases, too, it is not always necessary that one be true and the other false. For when Socrates exists (ὄντος μὲν γὰρ Σωκράτους)[43] one will be true and the other will be false, but when Socrates does not exist both will be false, for neither "Socrates is ill" nor "Socrates is well" will be true when Socrates does not exist at all. In the case of privation and possession, when the subject does not exist at all neither proposition will be true, but when the subject does exist it is not always the case that one or another of the opposing predications is true. For "Socrates has sight" is opposed to "Socrates is blind" as privation and disposition, and even though Socrates is alive it is not necessary that one or the other be true or false. For when he is not yet by nature such as to have sight, both propositions are false. But when Socrates does not exist at all, both "he has sight" and "he is blind" are false. In the case of affirmation and negation, whether the subject exists or not, one must be false and the other true. For in the case of "Socrates is ill" and "it is not the case that Socrates is ill," if Socrates does not exist it is clear that one or the other is true or false, and likewise if Socrates does not exist, for if he does not exist "Socrates is ill" is false and "it is not the case that Socrates is ill" is true. (*Cat.* 10.13a37–b3, 13b12–35)

Aristotle is here concerned with the logical relations between opposed predications in the case in which the subject of the predication exists and in the case in which it does not. The examples Aristotle presents are all of *singular* predications (positive and negative predications of Socrates), but there is no indication that Aristotle's remarks are to be

---

43. The verb εἶναι used in an absolute sense, when said of a person or animal, often has the sense of "to be alive." (See Kahn on the "vital" sense of "to be," in *Verb "Be,"* 240–45.) It clearly has this sense in this passage. But Aristotle is here making a logical point concerning the truth or falsity of propositions with nonexistent subjects; he is not simply concerned with which terms can be predicated of subject terms denoting dead or inorganic entities. "Socrates lives" is here taken as equivalent to "Socrates exists." For another view, see W. Jacobs, "Aristotle on Nonreferring Subjects," *Phronesis* 24 (1979): 282–300. I discuss Jacobs' interpretation of this passage in n. 45.

taken as applicable to this kind of predication alone, for Aristotle often considers predicating something of a universal subject to be a kind of simple predication. We can therefore take Aristotle's remarks here as having bearing on the semantic analysis of the universal affirmations employed in demonstrations. Aristotle has earlier identified some predicates that are paired to a contrary such that "in the case of those subjects in which they naturally come about or of necessity one or the other belongs, there is no middle between them" (*Cat.* 10.11b38–12a2).[44] His point here is that, if a subject does not exist, the predication of either contrary is false.[45] (The case is similar with terms opposed as privation and disposition, except that in this latter case there is another possibility in which the predication of either contrary is false: the subject is not yet old enough to have the disposition.) An example may help here. "Animal" and "not-an-animal" are contraries (for every subject falling under the category of substance is either an animal or not an animal). Given that there are no goatstags, both "a goatstag is an animal" and "a goatstag is not-an-animal" are false. It follows that the contradictory of both predications ("It is not the case that a goatstag is an animal" and "It is not the case that a goatstag is not-an-animal") are true. But these contradictories do not have the form that is necessary if either is to serve as a premise of a demonstration of an affirmative conclusion regarding goatstags. This is why there will be no demonstrations by which the attributes of goatstags are explained.[46]

44. At *Cat.* 10.12a4–9 health and disease, the examples in the present passage, are identified as such contraries.

45. On the basis of *DI* 6.17a20–26 and 10.19b5–12 Jacobs ("Aristotle," 282) argues that Aristotle does "not accept the possible referential failure of the subject of an assertion." He offers an interpretation of *Cat.* 10, according to which the absolute use of εἶναι in this chapter is in accordance with the vital sense of the verb alone. According to Jacobs, the whole passage concerns the issue of the applicability of certain predicates to certain subjects. In my opinion the interpretation Jacobs presents of this passage is somewhat strained. He asserts that the unstated subject of εἶναι at *Cat.* 10.13b20–21 and 28 (where it appears that Aristotle is making a point about all assertions) is Σωκράτης and that the understood predicate complement of the forms of εἶναι in 13b20–27 is ζῷον, which term does not appear in the text. I suggest that although Jacobs' interpretation of the passages of *DI* concerning the nature of assertions may be correct, the view Aristotle expresses there is not sophisticated enough to explain how there can be any meaningful assertions regarding nonexistent entities. The thesis of *Cat.* 10 (that these assertions are false) and the discussion of nominal definitions within *APo.* are first steps toward developing a semantic account able to explain this.

46. There are several important respects in which Aristotle's account of the semantics

Aristotle is not wont to engage in semantics for its own sake. His analyses of language are often meant to have bearing on the nature of things, of which language speaks. The present case is no exception. The fundamental insight of Aristotle's analysis is that no proposition with a determinate subject is true simply by virtue of the formal properties of the language being employed. Rather, the truth of any proposition concerning some kind A must be grounded in a consideration of A's as they really are[47] (or at least as they are considered within a certain context).[48] This is in contrast to the truth of a proposition such as $\forall x(Ax \rightarrow$

---

of propositions whose subjects do not exist parallels Russell's theory of definite descriptions (meant to solve a problem different from that which is Aristotle's concern). First, both Aristotle and Russell assert that propositions whose subject terms do not denote any existing thing are false, not nonsensical. See B. Russell, *Essays in Analysis*, ed. D. Lackey (London: Allen and Unwin, 1973), 108. (Russell, however, does not deny that propositions whose subjects are true names, not abbreviated descriptions, must have referring subjects in order to be meaningful, although few if any such names would be employed in ordinary language.) Second, both Aristotle and Russell recognize that there are two kinds of negation of a positive predication of a nonexistent subject. When the whole proposition is negated, the negation is true, while, when the predication alone is negated, the negation is false. (See ibid., 116). But a Russellian description holds only of a particular, not of a universal. Aristotle, unlike Russell, focuses on the problem of presenting a semantic analysis of "All A's are B's" when there is no A. If A is considered a kind that constitutes a single subject, Aristotle must assert that "All A's are B's" is *false* if there are no A's, just as a proposition with a particular subject is false if that subject does not exist. (Russell, of course, would translate "All A's are B's" as $\forall x\ (Ax \rightarrow Bx)$, which, given $\sim\exists xAx$, is true no matter what B is, given the meaning of "$\rightarrow$.")

47. This is a main point of Ferejohn, *Origins*.

48. I make this qualification because, as Ackrill (*Aristotle's "Categories" and "De Interpretatione," Translated with Notes* [Oxford: Clarendon Press, 1963], 110–11) points out, it is not clear how one is to reconcile the dependence of the truth of a proposition such as "Socrates is well" on that of "Socrates exists" with Aristotle's assertion at *DI* 11.21a25–28, in which he says that from "Homer is a poet" we cannot infer "Homer is," on the grounds that in the first proposition "is" is predicated accidentally. I hold that the problem is to be resolved by distinguishing the two different kinds of discourses in which "Homer is a poet" and "Homer is" will find a place. We say "Homer is a poet" when we are speaking of literature and literary figures. In *this* context it *would* be proper to say "Homer is," thus distinguishing him from fictional authors such as Gustav Aschenbach. But we would not thereby be saying that Homer is in respect to his real essence (that of being a rational animal or a human being).

An alternative way of dealing with the problem is suggested to me by some comments of Sarah Broadie. She raises the possibility that Aristotle's requirement that the subject of propositions refer is a special feature of the language employed in scientific discourse, on account of the fact that such discourse is limited to revealing real causal connections among essences. On this view ordinary language may not necessarily conform to such a requirement.

Bx) & ~∃x(Ax), which would be true, no matter how things in the world happen to be.[49]

But Aristotle has no objection to purely stipulative definitions. No impropriety is committed in stipulating that a term *A* can be predicated of anything of which C, D, and E are predicated (or in presenting a nominal definition that reports that "C and D and E" is the meaning of *A* in some given language), even if there is nothing having all three attributes. We can consider such a definition a proposition whose subject is a linguistic entity, *A*; we are predicating of that term its use in language. We are *not* predicating anything of A's, not even that they are C's, D's, and E's, for, as we have seen, such a predication would be false given that there are no A's. This is why Aristotle writes:

> One who knows what a human being or anything else is must also know that it exists (ὅτι ἔστιν), for nobody knows what a nonexistent thing is, but, while you know what the phrase or word means (τί . . . σημαίνει ὁ λόγος ἢ τὸ ὄνομα) when I say "goatstag," it is impossible to know what a goatstag is. (*APo.* 2.7.92b4–8)

Aristotle is here distinguishing accounts of the meaning of a term from accounts of the τί ἐστι, that is, definitions identifying the essence of a real thing denoted by a term.[50] If a term has no reference, only the first kind of definition of it can be an object of knowledge,[51] for a latter

49. As Sarah Broadie has pointed out to me, because the truth of any proposition that is an object of science rests on the existence of the subject, and the existence of that subject must be grounded on *some* empirical evidence, that proposition would not be regarded by Aristotle as "logically or analytically necessary."

50. This is the same as the distinction made at 2.7.92b27–28, 10.93b29–35, 94a8–14, between a nominal definition and the sort of definition that serves as a scientific first principle. I also take this distinction to be alluded to in the contrast Aristotle makes between an assumption of τοδὶ εἶναι (1.10.76b6) and of τί σημαίνει (1.10.76b7, 15). Of the two assumptions only the first sort, the assumption of the essence of a basic kind studied by a science, has the status of being a scientific first principle.

51. R. Bolton ("Essentialism and Semantic Theory in Aristotle's *Posterior Analytics* II.7–10," *Philosophical Review* 85 [1976]: 524) argues that within *APo.* Aristotle presents a theory of natural kind terms according to which any proposition has meaning insofar as its terms refer to those real entities that would be recognized by certain authorities as instantiating the kind in question. One of Bolton's key arguments for this thesis rests on the claim that Aristotle never considers a nominal definition (the sort of definition discussed at *APo.* 2.10.93b29–35) as merely expressing the *meaning* of a term; it must always "focus on" an essence of a real kind. But that there can be nominal definitions of nonreferring terms is clearly indicated by the fact that Aristotle's description of the sort of account there can be of a term such as *goatstag* (τί σημαίνει ὁ λόγος ἢ τὸ ὄνομα) in

kind of definition, expressing the essence of something that does not exist, would be false and hence not an object of knowledge.[52]

It should be noted that a nominal definition such as "a goatstag is an animal with the head of a goat and the body of a stag" is indistinguishable from a real definition that would be given of goatstags if there were such things. So, there is a need for some device that distinguishes the two kinds of definitions in order to prevent inferences regarding kinds from mere nominal definitions. We do not want "all goatstags have a backbone," for example, to be among the demonstrable propositions of zoology, for this proposition is false.

Two devices could be employed for this. Some definitions could be rephrased so as to make explicit the fact that they are merely reporting the way in which terms are employed. Or we could have for each science a set of first principles stating which basic kinds exist as subject matter for it and which do not.[53] These principles, I propose, are Aristotelian assumptions of existence.

We can now see how existence claims can act as starting points of demonstration without acting as premises of demonstration. Any demonstrated proposition presupposes the existence of its subject. So, in order to insure that the inferences we make on the basis of definitions are sound, the existence of the subjects of the sciences needs to be explicitly posited.[54] To explicitly express such an existence claim as a

---

2.7.92b6–7 is essentially the same as the more elaborate description of the first kind of scientific definition in 2.10.93b30–32: λόγος τοῦ τί σημαίνει τὸ ὄνομα ἢ λόγος ἕτερος ὀνοματώδης, οἷον τί σημαίνει τρίγωνον.

52. In his commentary (L.II,l.vi,461) Aquinas apparently presents Aristotle's objection as purely epistemological: if a thing is not given in nature, there is no way in which we can grasp what that thing is. This is in line with the metaphysics of Aquinas, according to which an act of existing is responsible for an essence's being an essence of something that exists. For Aquinas there could very well be an essence of a goatstag, although no human being could know it or express it in a definition. But I have here drawn attention to views Aristotle has on language and its relation to reality that show that Aristotle could not have shared Aquinas' views on the relation between essence and existence. For Aristotle an essence must be an essence of something real, for, otherwise, the definition expressing that essence would be false.

53. At *APo.* 1.2.72a20 a statement of τὸ μὴ εἶναι τι, as well as a statement of τὸ εἶναι τι, is called a hypothesis. I take such negative hypotheses to have the function of banishing all talk of nonexistent subjects from the inquiry and exposition of a science. Aristotle's denial of the void at *Phys.* 4.8.216b20–21 is an example of such a negative hypothesis.

54. Cf. S. Mansion, *Le jugement d'existence chez Aristote*, 2d ed. (Louvain: Publications Universitaires de Louvain, 1976), 181: "La définition est une des pièces maîtresses de la science et comme la science a pour object les êtres réels, la définition doit porter sur les choses existantes et en exprimer l'essence." I cannot, however, agree with Mansion's

first principle is to say that these are basic subjects, of which there is no other cause. For, otherwise, the existence of the subject would be expressed as a conclusion of a demonstration.

There are three areas in which the assumption of the existence of a kind will have special importance. First, ancient travelers would often bring back tales of unknown creatures and minerals with fantastic attributes. The Lyceum took an interest in these stories, as the Aristotelian work *On Marvelous Things Heard* testifies. There must have been a presumption that among all such stories some must have been true and of value for future researchers. Members of the Lyceum were to keep these questions in mind so that future observation could determine whether the things that were described really exist and whether the sciences are to concern themselves with them. Aristotle's use of "goatstag" (2.7.92b7) and "centaur" (2.1.89b32) to illustrate subjects of which we ask εἰ ἔστι confirms the importance Aristotle placed on ascertaining the existence or nonexistence of exotic things.

Second, there are theoretical entities of which it is not immediately clear whether or not they are proper subjects for the sciences. In the *Physics* Aristotle claims that time and place are proper objects of study for the physicist. Because there is a sense in which there are such things, their properties are to be explicated and explained. We have also seen

claim that without an assertion of existence a definition would be a mere tautology (156, 328 n. 53). For, if we consider a definition of a subject that we do not know to exist as a nominal definition, it is properly considered as predicating of some *term* what that term means (cf. 2.7.92b5–8); this will be true but not necessarily so. (Different dialects sometimes give the same term different meanings.) If this definition is considered as *stipulative*, it lies outside of the discourse employing the term and, hence, does not admit a truth value.

The role I have given to existence claims in the sciences is also much like that given by Ferejohn "Definition." Ferejohn, too, sees the ascertaining of the existence of a genus as necessary for endorsing the predications given in definitions so that they may be employed as premises of demonstrations. We differ on exactly how this is so. Ferejohn suggests that, after hypothesizing the existence of a genus, we are entitled to employ the method of division in respect to that genus and thus generate definitions of the species. These definitions are in turn to ground the primitive premises of demonstration (393). But Ferejohn gives the method of division too great a role in the Aristotelian scheme of the sciences. In *APo.* 1.31 and 2.5 Aristotle makes it clear that one can neither learn nor prove anything new through division; the method, rather, serves to ensure that definitions are properly ordered and complete. Any true definitional predication of a thing must already be recognized as true and definitional prior to the employment of the method (*APo.* 2.5.91b18–26). Hence, it cannot be the method of division as such that allows us to recognize that a certain predication is part of what it is to be of some existent kind.

that he denies this of the void. Such theoretical subjects will be in special need of having their existence explicitly hypothesized, since they cannot be directly perceived as can real substances. This is why at 1.10.76b16–18 Aristotle indicates that some subjects need to have their existence explicitly hypothesized, while others do not. Even though subjects like the hot and the cold are not substantial, they are readily apparent to the senses, so that there is no question of their being in the world and hence of their being fit objects for study. This is not the case for a subject such as "number."[55]

The third area in which the hypothesizing of existence takes on a special importance is unique to mathematics. In geometrical proofs the possibility of some constructions is taken for granted, as in postulates 1–4 of the first book of Euclid's *Elements*. I have argued above that for Aristotle this is, in effect, the hypothesizing of the existence of certain geometrical entities.

F. Hintikka on Existence Claims in the Sciences

Hintikka has put forward another account of the role of existence claims in the sciences. As I do, Hintikka rejects the efforts of Gomez-Lobo to banish assertions of existence from the sciences as conceived by Aristotle. He correctly points out that εἶναι is often used existentially by Aristotle, especially in the *Posterior Analytics*. He also points out that this use, as well as the use of the "is" of identity and the "is" of predication, would not be taken as a *distinct* sense of εἶναι, since Aristotle was oblivious of the Frege/Russell treatment of these as distinct senses of "to be."[56]

Hintikka's account of the role of assertions of existence is as follows. Given a genus, divided into all of its subordinate genera and species,

---

55. G.E.L. Owen ("Aristotle on the Snares of Ontology," in *New Essays on Plato and Aristotle*, ed. R. Bambrough [New York: Humanities Press, 1965], 90–92; reprinted in *Logic, Science, and Dialectic*, ed. M.C. Nussbaum [Ithaca, N.Y.: Cornell University Press, 1986]) shows that, in the case of theoretically questionable entities (such as time and the infinite), the task of showing that they exist is inextricably bound up with the tasks of showing what they are and how they exist.

56. J. Hintikka, "The Varieties of Being in Aristotle," in *The Logic of Being*, ed. S. Knuuttila and J. Hintikka (Dordrecht, Neth.: D. Reidel, 1986), 81–92. Hintikka argues that, in failing to make the Frege/Russell distinction, Aristotle is in a better position to grasp the semantics of natural language than those semanticists who take the predicate calculus as a model language.

the existence of any of these subclasses can be demonstrated given the existence of the highest genus. For example, suppose that the existence of the highest genus G is assumed and that a species A is defined as falling under G. Then, treating existence as a monadic predicate, the following demonstration is possible: Every G exists, every A is a G; hence, every A exists. In such a case the only assertion of existence that is required is that regarding the highest genus; every other would be superfluous, since the existence of every subordinate genus and species is demonstrable.[57]

Hintikka identifies these highest genera with the genera that constitute the subject matter of the sciences, which, at *Posterior Analytics* 1.10.76b2–5, Aristotle says are assumed to exist. Since such genera cannot, within this science, be defined by giving the genus and differentia, a definition of the genus can be only a useless tautology. The true role of a definition of the highest genus studied by a science is to assert the existence of the genus. The καθ' αὑτὰ πάθη discussed at 1.10.76b6–11 are taken by Hintikka to be the species and subgenera, whose existence must be demonstrated on the basis of the existence of the highest genus.[58]

For Hintikka, then, assertions of existence are given in certain definitions; there are no Aristotelian first principles having the linguistic form "x exists." Hintikka can thus interpret 1.2.72a14–24 with ease; by ὁρισμός (as well as τί ἐστι) here, Aristotle refers to a nominal definition, which can play no role as a premise of demonstration. A hypothesis, said to assume that the subject is x or is not x, is any definitional first principle, whether of the highest genera (and therefore with existential import) or not, that can be used as a demonstrative premise.[59] Hintikka says neither what role he takes nominal definitions to have in scientific inquiry nor what relation he thinks they have to hypotheses.

This ingenious theory has two major flaws, which, I believe, are cause enough to reject it.

First, although Hintikka cites *Posterior Analytics* 1.10.76a31–36, 76b2–16, as supporting his interpretation, he omits mention of the examples Aristotle gives in these passages: arithmetic assumes the existence of the monad (76a34, 76b4) and demonstrates the existence of

57. Hintikka, "On the Ingredients," 62–66; "Varieties of Being," 87–88.
58. Hintikka, "On the Ingredients," 64.
59. Ibid., 66–69.

odd, even, rectangular, and cube (76b7–8), while geometry assumes the existence of magnitude (76a36) and of points and lines (76b5) and demonstrates the existence of irrational, inflection, and verging (76b9). Every science will have its own genus; that of geometry is magnitude. Why then does Aristotle say that geometry assumes the existence of two subordinate genera: points and lines? And are we to take odd, even, rectangular, and cube as falling under the genus monad? Aristotle's examples belie Hintikka's interpretation.[60]

Second, the sort of demonstration that Hintikka believes Aristotle would employ to secure the existence of a species would be worthless in the actual practice of a science in determining whether or not it exists. Suppose a zoologist were engaged in the inquiry to determine whether or not centaurs exist (cf. 2.1.89a32). As Hintikka would have it, the matter could be easily resolved by means of a demonstration: animals exist, centaurs are animals; hence centaurs exist. But this obviously will not do, for centaurs do not exist. If centaurs did exist, we would know this by direct observation or from some reputable authority. Their existence would be expressed within the exposition of a science by the explicit assertion "centaurs exist."

One who wished to press Hintikka's interpretation might object that "centaurs are animals" could not be employed as a premise in the demonstration of the existence of centaurs, for the reason that there can be nominal definitions alone of nonexistents (2.7.92b5–8) and, as Hintikka points out, nominal definitions "have no assertive force"[61] and hence cannot be employed to generate premises of demonstration. But how is such a nominal definition to be distinguished from a real definition, if not by the assertion that there really is an entity whose essence is that given in the definition?

I conclude that we are on sure ground in taking Aristotle's view to be that the existence of the epistemic substances is to be assumed and that of the καθ' αὑτὰ συμβεβηκότα is to be demonstrated.

60. This point is made by D. Frede ("Comment on Hintikka's Paper 'On the Ingredients of an Aristotelian Science,'" *Synthese* 28 [1974]: 82). Frede holds that the genera alone are explicitly hypothesized to exist and that only the definitions of such genera are immediately such as to be employed in demonstrations. In this we agree. But she does not make clear that Aristotle is using γένος in a loose sense in 1.10.76b13, according to which any species is a genus. If this is kept in mind, Hintikka's point that definitions of such genera alone provide premises for few demonstrations of scientific interest loses its force. J. Hintikka, "Reply to Dorothea Frede," *Synthese* 28 (1974): 91.

61. Hintikka, "On the Ingredients," 67.

G. Epistemic Substances

We see that within the first book of the *Posterior Analytics* Aristotle distinguishes two different kinds of entities studied by the science, those whose existence is assumed by the science studying them and those whose existence is demonstrated by this science. It is clear that substantial kinds will fall into the first class. We have seen that, because the demonstration of the existence of a thing, as Aristotle understands it, can come about only through the demonstration that the thing inheres in some more basic substrate, the existence of the ultimate substrates, the substances, will be indemonstrable. These substantial kinds will be among the genera posited by the sciences.[62] But the mathematical examples Aristotle presents in 1.2 and 10 make clear that substantial kinds are not the only genera whose existence is posited by the sciences, and thus not all metaphysical accidents are such as to have their existence demonstrated by the sciences studying them. There is no middle term by which a mathematician can demonstrate that there are such things as points, lines, or monads; as we have seen, these mathematical primitives are taken by Aristotle to constitute the genera of the sciences studying them. If mathematical primitives inhere in certain substances, why cannot some feature of those substances serve as a middle term, on the basis of which the existence of such primitives can be demonstrated? The answer is that, although there is an *ontological* dependence of the mathematical primitive on some material substance, the causal link between the substantial kind and the mathematical property of the kind cannot be made manifest through any kind of demonstration. One cannot demonstrate that mathematical attributes exist on the basis of the definitions of composite substances. For this reason, no nonmathematical feature of a substance in which a quantitative attribute inheres can serve as a middle term on the basis of which that mathematical attribute's existence can be demonstrated.

We see that, according to Aristotle's philosophy of scientific explanation, there are certain kinds that are not substantial, insofar as they are metaphysically dependent on some substance in which they are to inhere but are nonetheless to be considered as ultimate substrates by

62. The argument of *APo.* 1.22.83a1–83b31 rests on the assumption that there are ultimate subjects of the predications that are the objects of scientific understanding. See Ross, *Aristotle's "Prior and Posterior Analytics,"* 577.

the sciences studying them.[63] There are several passages in which Aristotle shows himself to be aware of these implications of his theory of demonstration.

In *Posterior Analytics* 1.9 Aristotle asserts that each science rests upon its own set of indemonstrable principles. For Aristotle mathematical predications can be explained only on the basis of mathematical first principles. The definitions on which mathematical demonstrations rest will be definitions of only mathematical entities. Accordingly, the mathematician considers what is not really an ultimate metaphysical substrate as an ultimate substrate of predication. Aristotle presents his most explicit account of how this is so in *Metaphysics* 13.3, in which it is explained that, although mathematical entities are from the point of view of metaphysics accidents of substances, the mathematician "abstracts" the fact that they are predicated of some substance or other from his consideration of them.[64] These entities are considered insofar as they are quantitative, not insofar as they are attributes of some substance.[65]

---

63. On this, see ibid., 629, 633. Within *APo.* all examples of such kinds are mathematical, but it is possible that there would be other nonsubstantial kinds that Aristotle would believe are taken to be primitive by the science studying them.

64. Aristotle makes clear in *Metaph.* 11.3.1061a28–b3 that the process of mathematical abstraction is not one whereby the attribute being considered is "stripped off" of the whole entity in which it inheres; rather, it is those attributes of the entity that are not logically entailed by the attribute in question that are "stripped off." See I. Mueller, "Aristotle on Geometrical Objects," in *Archiv für Geschichte der Philosophie* 52 (1970): 160–62; reprinted in *Articles on Aristotle*, vol. 3: *Metaphysics*, ed. J. Barnes, M. Schofield, and R. Sorabji (London: Duckworth, 1979); and E. Hussey, *Aristotle's "Physics," books III and IV, translated with Notes* (Oxford: Clarendon Press, 1991), 113–25.

65. In *Metaph.* 13.3 Aristotle asserts that the mathematician considers mathematical entities insofar as they are mathematical, which is why disregarding the fact that they inhere in nonmathematical entities entails no falsehood. The term translated as "insofar as" is ᾗ, equivalent to the Latin *qua*. As A. Bäck (*On Reduplication* [Leiden, Neth.: E.J. Brill, forthcoming], chap., 18, sec. 1) has argued, in the proposition "A is B ᾗ C" (taken as "an essential *qua* predication") the clause "ᾗ C" "is attached to the predicate, and not to the subject." It is an adverb, identifying C as a middle term by virtue of which "A is B" can be deduced. In the context of scientific explanation, the ᾗ clause will identify that whose essence serves to explain the conclusion of an explanatory demonstration. (This is why at *APo.* 1.4.73b28–29 Aristotle writes, "the phrases καθ' αὑτό and ᾗ αὑτό mean the same thing" [τὸ καθ' αὑτὸ δὲ καὶ ᾗ αὑτὸ ταὐτόν]. The phrase ᾗ x is synonymous with καθ' ὃ ἐστι x, in the sense of καθ' ὅ discussed at *Metaph.* 5.18.1022a14–18, in which the ὅ identifies that essence that is the explanatory ground of the predication of x of some subject in which x inheres.) To consider A ᾗ C is to take C as that whose essence is responsible for the attributes of A. This involves disregarding all of those attributes of A that are not to be explained on the basis of C.

Although there is no evidence that Aristotle's account of mathematical abstraction had been developed by the time of the writing of the *Posterior Analytics*, the following passage makes clear that the doctrine was already implicit:

> Mathematics deals with kinds (εἴδη), for they are not predicated of some subject (οὐ γὰρ καθ' ὑποκειμένου τινός). For even if geometrical entities are predicated of some subject, still geometry does not study them insofar as they are predicated of a subject. (*APo.* 1.13.79a7–10)

This passage, in which Aristotle identifies certain kinds (εἴδη) as the subject matter of mathematics, is not an admission of Platonism; Aristotle grants that kinds are, as a matter of fact, always predicated of some substrate. All that Aristotle is saying is that the mathematical sciences do not take account of the fact that certain mathematical kinds, those that ground the explanation of other mathematical attributes,[66] are predicated of some substrate. *From the point of view of mathematics*, they are considered as predicated of no substrate, that is, as substantial.[67] As such, the mathematician will consider these mathematical primitives as having no cause other than their own essences. This may involve some distortion, for the predication of a mathematical entity of some subject may well be such as can be rendered intelligible on the basis of the essence of that subject. (E.g., presumably a botanist could explain the sphericity of a grapefruit.) But even if such explanation were possible, it would not be within the province of the mathematician to provide it.

---

66. Aristotle is never clear on exactly which mathematical entities are the substrates for others. In arithmetic the basic entity is sometimes taken to be the monad (*APo.* 1.1.71a15; 1.2.72a21–24; 10.76a34–35, 76b4–5) and sometimes number (1.7.75b4–5; 1.10.75b2, 18; 1.33.88b28). In geometry the basic entities are sometimes taken to be points and lines (1.7.75b17; 10.75b5; 27.87a36) and sometimes magnitude (1.7.75b4–5; 10.76a35–36, 76b1; 32.88b29). See Ross, *Aristotle's "Prior and Posterior Analytics,"* 539; and sec. B.

67. See Ross, *Aristotle's "Prior and Posterior Analytics,"* 633. Such an object, which is not substantial but is considered as substantial, will belong to the class of what Hussey calls "representative objects," of which it is true "by definition that they possess just those properties which (i) are shared by all (actual or possible) individual members of the class they represent, and (ii) are *representative* properties, i.e., belong to the individuals qua members of that class" (*Aristotle's "Physics,"* 118). This means that any attributes these objects have that are consequent to their being predicated of some more basic substrate are ignored. Hussey does not limit representative objects to the mathematical primitives.

Nowhere in the *Posterior Analytics* does Aristotle show any sign that the logic of mathematical explanation is in any way different from the logic of explanations belonging to other sciences. Indeed, examples of purported mathematical demonstrations abound, even though, notoriously, they cannot be cast as first-figure universal syllogisms, the logical form of most Aristotelian demonstrations. It is reasonable to infer that, for Aristotle, what distinguishes all of the theoretical primitives of all of the sciences, not just the mathematical sciences, is the fact that they are considered by the sciences to be predicated of no more fundamental substrate.

Accordingly, scientific understanding (ἐπιστήμη) involves positing certain kinds as ultimate substrates, regardless of their real metaphysical status. Within the context of this sort of discourse, the definitions of these kinds serve as a foundation on which certain attributes of these kinds must rest. Like real substances, some of these kinds have certain essences. Unlike real substances, these kinds are ultimate substrates only from the point of view of the ἐπιστήμη dealing with them. This is why I choose to refer to them as *epistemic substances*.[68] Because, from the perspective of the sciences studying them, their existence is not to be explained on the basis of any substrate in which they inhere; they are those things "of which there is no different cause."

Some nonsubstantial attributes of substances can be considered to be epistemic subjects by taking a point of view that disregards the fact that they do in fact inhere in some substrate. This point of view will be that of whatever science it is that studies them. For example, the light reflecting off of Socrates' forehead must be regarded as an accident of Socrates if one adopts the point of view of biology or ethics, but it is an epistemically substantial entity from the point of view of optics, which disregards the fact that the light has a source of this or that substantial nature. An accident can be abstracted in such a way as to be taken to be epistemically substantial only if it is considered by means of the proper science.[69]

68. This phrase was suggested to me by Sarah Broadie.

69. Whether or not an accident has the status of being καθ' αὑτὸ συμβεβηκός will similarly be determined by the perspective of the science one is adopting. The sum of the interior angles of a triangular piece of wood is a καθ' αὑτὸ συμβεβηκός attribute from the point of view of geometry, which posits as an epistemic substance that to which that attribute is καθ' αὑτὸ συμβεβηκός, but not from the point of view of chemistry, since the attribute's existence cannot be explained on the basis of a consideration of wood as such.

But not every attribute that can be conceptualized as an ultimate substrate of predication is epistemically substantial. An epistemically substantial attribute must have some science that takes it as its subject matter. For example, it is possible for one to consider the attributes of eclipses and thunder without attending to the subjects of which eclipses and thunder are predicated. But doing so precludes adequate explanation of their attributes. For example, an adequate scientific explanation of thunder requires the theoretical support afforded by physics and acoustics, which take as subjects such things as the elements and moving air, the subjects in which the thunder truly inheres. This is not merely a feature of the way in which the sciences have been classified. It is, rather, on account of the nature of thunder itself that its attributes are to be made intelligible only if one takes it as predicated of the theoretical primitives of meteorology and acoustics.

So, the fact that an attribute can be considered an ultimate subject of predication is not a sufficient condition for its being an epistemically substantial attribute. An epistemic substance must have a simple essence that can be independently grasped. Its essence must be expressible in a definition grounding explanatory demonstrations of its attributes. The essence of such an attribute acts as a causal and explanatory "bottom" of all of its καθ' αὑτὰ συμβεβηκότα attributes, just as a substance acts as the ontological bottom of all of its attributes.

Further evidence for epistemic substances in Aristotle's thought is found in *Metaphysics* 5.18.1022a32–35. Aristotle here distinguishes a sense of καθ' αὑτό according to which a thing (not a predication) can be called καθ' αὑτό.

> Further, that of which there is no different cause is καθ' αὑτό. For there are many causes of "human being": "animal," "two-footed," but nonetheless the human being is a human being on account of itself.
>
> ἔτι οὗ μὴ ἔστιν ἄλλο αἴτιον· τοῦ γὰρ ἀνθρώπου πολλὰ αἴτια, τὸ ζῷον, τὸ δίπουν, ἀλλ' ὅμω ς καθ' αὑτὸν ἄνθρω πος ὁ ἄνθρω πος ἐστιν.

Aristotle is here appealing to the "primary" sense of καθ' ὁ delineated at *Metaphysics* 5.18.1022a14–18, according to which a predicate belongs to a subject κατά, the formal or essential cause of the predication. Those entities that are καθ' αὑτά in this sense are those that are what they are on account of no formal or essential cause other than their very own form. The predicates presented in their definition are their formal or essential causes, but since they constitute a true unity they do not signify anything ontologically different from themselves.

Entities whose essences are true unities will not have their formal or essential cause in some more basic substrate in which they inhere, for, if they did, their formal cause *would* be something other than themselves. Hence, they comprehend those entities that are either substantial or are considered substantial by those sciences that are concerned with their essences. What Aristotle here calls things "of which there is no different cause" are epistemic substances.[70]

I have shown that Aristotle does distinguish between the epistemic substances and the other entities studied by a science. I have also shown that the former class is coextensive with the class of entities whose existence is assumed. And, finally, I have shown that the class of those entities that are studied by a science but are not epistemically substantial is coextensive with the class of those entities studied by a science whose existence is somehow to be demonstrated.

70. The sense of καθ' αὑτό Aristotle is delineating here is closely related to the third of the four senses of καθ' αὑτό delineated in *APo.* 1.4: "In another sense, what is not said of some other subject is καθ' αὑτό. For example, that which is walking is walking, as being something different (οἷον τὸ βαδίζον τερόν τι ὂν βαδίζον ἐστι), and what is white is, considered as white, something different as well. But substance (οὐσία) and all of those terms that signify some 'this' (τόδε τι) are, considered as being the very things they are, not something different. So I call those things that are not predicated of a subject καθ' αὑτά, and those that are predicated of a subject I call accidents" (1.4.73b5–10). It is generally agreed that Aristotle is here distinguishing substance from accidents on the grounds of whether or not a thing is predicated of some subject ontologically different from it (see, for example, Ross [*Aristotle's "Prior and Posterior Analytics,"* 519]; and Barnes [*Aristotle's "Posterior Analytics,"* 115–18]). The class of epistemic substances will include this class but will also include those things, such as mathematical primitives, that do inhere in substrates ontologically different from themselves but are not considered as so inherent by the sciences studying them. McKirahan (*Principles and Proofs*, 93–94) interprets this passage as referring to the subjects of the sciences, not to real metaphysical subjects. This is an appealing interpretation, since it integrates Aristotle's point into the demonstrative theory of the *APo.* Unfortunately, it is not required by the text itself. Further, McKirahan does not say whether what he calls "derivative subjects" (which I consider attributes [see n. 15] are καθ' αὑτά in this way. Clearly, they are not, since their existence is caused by that of the primitive subjects.

4
The Aporiai

It is the thesis of the present study that within *Posterior Analytics* 2.8–10 Aristotle sketches the logical structure of a demonstration of a καθ' αὑτὸ συμβεβηκός predication. But within these chapters Aristotle never explicitly asserts that this is the problem with which he is concerned. Rather, 2.8–10 follow a series of aporiai, or conceptual puzzles (rivaled for length only by the series of aporiai of *Metaph.* 3), dealing with the issue of whether one can know (εἰδέναι) the same thing through demonstration and definition (2.3.93b2–3). Anyone with even a casual familiarity with the theory of explanation put forward in the first book of the *Posterior Analytics* would be surprised that Aristotle would at this point even bring up the issue, let alone regard it as generating a sequence of serious problems. For a definition is a first principle (*APo.* 1.2.72a14–21). A demonstration is a deduction that rests upon first principles, which are themselves indemonstrable (*APo.* 2.3.72b18–25). The conclusion that one cannot demonstrate a definition seems inescapable.

Yet Aristotle devotes five chapters to mulling over various aporiai involved with the notion of demonstrating a definition and then proceeds to indicate how a definition could, in a sense, be demonstrated. As we shall see in the next chapter, on the interpretation that has been most often favored, within *Posterior Analytics* 2.8–10 Aristotle shows how part of a definition can be a conclusion of a certain kind of *dialectical* syllogism, not to be considered demonstrative. We shall see how this is a problematic interpretation of the text. But its implausibility is also evident simply on the grounds that it would have Aristotle atypically making much ado about nothing. On the interpretation I reject, the aporiai that Aristotle spins at great length would contribute nothing to the resolution of what turns out to be a pseudoproblem.

As has often been remarked in regard to other contexts, the setting out of Aristotelian aporiai has a crucial dialectical role to play within

Aristotle's own exposition of some subject.[1] The aporiai are not intended merely to confuse the reader prior to his or her being illuminated by Aristotle's exposition of the truth. Rather, they constitute the important problems that any adequate account needs to be able to solve. An adequate solution to problems can be solved only once those problems themselves are understood. Aristotle's exposition of philosophical aporiai are meant to make possible such understanding.

This chapter is devoted to considering the aporetic chapters *Posterior Analytics* 2.3–7. I shall show how a close examination of the problems Aristotle raises both reveals the nature of the philosophical question with which Aristotle is dealing and indicates the nature of his answer.[2]

A. *Posterior Analytics* 2.3

Posterior Analytics 2.3 opens as follows:

> Now it is clear that all of our inquiries concern the middle term. But let us relate how one reveals the "what is it" (πῶς δὲ τὸ τί ἐστι δείκνυται) and how one is led up to it (καὶ τίς ὁ τρόπος τῆς ἀναγωγῆς), and what a definition is, and of what things it is, first going through the aporiai about them. Let the starting point of what is to come be what is most appropriate to the following arguments.[3] (90a35–b1)

1. See especially G.E.L. Owen, "'Tithenai ta Phainomena,'" in *Aristote et les problèmes de méthode*, ed. S. Mansion (Louvain: Publications Universitaires de Louvain, 1961), 83–92; reprinted in *Articles on Aristotle*, vol. 1: *Science*, ed. J. Barnes, M. Schofield, and S. Sorabji (London: Duckworth, 1975); and in *Logic, Science, and Dialectic*, ed. M.C. Nussbaum (Ithaca, N.Y.: Cornell University Press, 1986).

2. Ferejohn, too, has pointed out that *APo.* 2.3–10 follows Aristotle's usual strategy of first presenting conceptual difficulties that result from a partial understanding of the matter at hand and then showing how the difficulties can be resolved on the basis of an adequate (i.e., Aristotelian) understanding of the matter. See M. Ferejohn, "Meno's Paradox and *De Re* Knowledge in Aristotle's Theory of Demonstration," *History of Philosophy Quarterly* 5 (1988): 111–12.

3. Barnes translates ἀρχὴ δ' ἔστω τῶν μελλόντων ἥπερ ἐστὶν οἰκειοτάτη τῶν ἐχομένων λόγων as "Let the start of what we are about to say be whatever is most appropriate to the neighboring arguments" (*Aristotle's "Posterior Analytics," Translated with Notes* [Oxford: Clarendon Press, 1975], 54) and asserts that "neighboring" (ἐχομένων) here must have the atypical sense of "preceding" (197). The problem with this is that the first considerations Aristotle presents derive from the theory of demonstration of *APo.* 1, not the discussion of the relationship between inquiry and demonstration of 2.1–2. Thus, Ross (*Aristotle's "Prior and Posterior Analytics"* [Oxford: Clarendon Press, 1949], 615) is right to take

Aristotle introduces his new discussion of definition by relating the issue of the nature of definition to the previous discussion, concerning the kinds of inquiry and the way in which they are resolved. He looks back to one of the main conclusions of 2.2: every inquiry has as its object the determination of the middle term of some explanatory demonstration. If we wish to know why P exists, we wish to know the essence of P. This is the case whether or not P is an epistemic substance. But it is unclear what the status of the definition of P would be in the case in which P is an attribute that is to be demonstrated as inherent in some subject. It is not to be a first principle; otherwise, the cause of the existence of P would already be assumed by the science studying P, and there would be no need to have a demonstration explaining why P inheres in some subject. Nor is it to be a demonstrated conclusion. For such a demonstration, like any other, would rest on an already ascertained definition of its major term, P.

At this point we should also recall that there are grave problems with the notion of an explanatory demonstration of a predication S is P (i.e., a demonstration that reveals the cause of the existence of P), if this predication is καθ' αὑτὸ συμβεβηκός (i.e., if it can be demonstrated to belong to its subject but is not implicitly definitional of that subject). These problems, discussed in chapter 1, are never explicitly presented by Aristotle but are implicit in the theory of demonstration as presented in *Posterior Analytics* 1.

We can therefore expect that the issue of the logical form of the demonstration of the existence of a καθ' αὑτὸ συμβεβηκός predicate P is relevant to the problem of the status of the definition of P within the science studying P, and vice versa. This is why Aristotle must again consider the nature and object of both definition and demonstration.

Aristotle considers three possibilities: (1) Every demonstrable proposition is expressible in a definition; (2) Every definable proposition is demonstrable; and (3) There are *some* propositions that are both demonstrable and expressible in a definition. Were possibility 1 the case, the problem of the demonstrability of the καθ' αὑτὰ συμβεβηκότα predications could be easily dismissed, since there would be no need to possess demonstrations of such propositions. Appeal could simply be

---

ἐχμένων to have its usual sense of "following." Aristotle's meaning is that his treatment of the relationship between definition and demonstration will have an appropriate starting point: a discussion of the basic roles definition and demonstration play in Aristotle's theory of explanation.

made to the relevant definition. Were possibility 2 the case, the problem of the status of the definitions of καθ' αὑτὰ συμβεβηκότα predicates could be dismissed, for every demonstration that S is P would simultaneously reveal what P is, and hence there would be no need for an independent definition of P. The middle term of such a demonstration would play the role of a definition of P: revealing what P is. Were possibility 3 the case, in those cases in which the expression of the essence of a καθ' αὑτὸ συμβεβηκός predicate is given in both a definition and a demonstration, one could dismiss either the problem of the possibility of a demonstration of its inherence in a subject or the problem of the status of its definition. In 2.3–4 Aristotle considers each of these possibilities in turn.

Aristotle needs to show first that there are demonstrable propositions that cannot be expressed in a definition. He begins by indicating that, while every definition is both universal and affirmative, not every syllogism has a universal and affirmative conclusion (90b3–7). This objection can be easily countered on the basis of *Posterior Analytics* 1. Not every syllogism is a demonstration, and the "better" demonstrations, those leading to scientific knowledge in the highest degree, are both universal and affirmative (*APo.* 1.24–25). Indeed, at *Posterior Analytics* 1.14.79a24–29 Aristotle argues for the importance of first-figure affirmative syllogisms, on the grounds that only those could serve as demonstrations helping us "hunt" for the essence of a thing, for a definition is a universal affirmative predication, and the only sort of demonstration that can have a universal affirmative conclusion is one that is a first-figure affirmative syllogism. Aristotle repeats this point at the beginning of 2.8 (93a6–9), which chapter is devoted to resolving the aporiai involved with the notion of a demonstration of a definition.

Aristotle next asserts that there will be universal affirmative propositions that are objects of the sciences but are not expressed in definitions. Hence, it is not the case that for every demonstrable proposition there is a definition expressing the same thing. As an example of a proposition that is demonstrable but not definitional, Aristotle presents his favorite example of a καθ' αὑτὸ συμβεβηκός predication: "every triangle is such that its angles are equal to the sum of two right angles" (90b7–17).

Aristotle appeals to the reader's awareness of the structure of actual scientific explanations. It is simply not the case that every object of scientific knowledge is part of an essence. Rather, the essence is the conjunction of all of the indemonstrable predications of a subject that are the ultimate premises for demonstrating other predications of the subject. If the theory of explanation presented in *Posterior Analytics* 1 is correct, there is a distinction between the indemonstrable principles and what is explained on the basis of those principles. The suggestion that every demonstrable predication is also expressed in a definition would obliterate this distinction.

Having shown that not every demonstrable predication is definitional, Aristotle proceeds to exclude possibility 2, showing that it is not the case that every definitional predication is demonstrable. He presents two arguments. He first restates the argument of 90b9–13: a demonstration of a definitional predication would be redundant, for a demonstration ought to impart knowledge one could not otherwise have (90b20–24). Aristotle also argues that the very possibility of demonstrations precludes an indefinite regress of demonstrative premises. This means that there must be indemonstrable principles of demonstrations, among which the definitions are numbered (90b24–27, cf. 1.3.72b18–25).

Aristotle then turns to considering possibility 3: whether it is conceivable that there are *some* predications that are the objects of both definition and demonstration.[4] This is the possibility he will ultimately advocate. In 2.8–10 he shows how a demonstration of a καθ' αὑτὸ συμβεβηκός predication serves as a kind of definition of the καθ' αὑτὸ συμβεβηκός predicate. But in 2.3.90b28–91a11 and in all of 2.4 Aristotle presents arguments to the effect that this is not the case. Like many of Aristotle's aporetic arguments, they are dialectical, and their worth consists in isolating the problem areas that any suitable account will need to address.

At 2.3.90b33–38 Aristotle states that within a demonstrable predication something is predicated of something else, while this is not the case for a definition. An Aristotelian definition is not to be regarded merely as a kind of analytic proposition, making the meanings of terms

4. As Barnes (*Aristotle's "Posterior Analytics,"* 198) points out, Aristotle does not directly argue that a proposition cannot be the object of both definition and demonstration; what Aristotle here denies is that a *single* demonstration can both serve to prove a predication and to identify an essence. This leaves open the possibility of a demonstration whose only object is a definition. But Aristotle has just twice argued that demonstration does not have a definition as its conclusion (2.3.90b9–13, 20–24). This is why Aristotle moves on to consider the possibility of a syllogism demonstrating *both* a καθ' αὑτὸ συμβεβηκός predication and an essence, in different ways.

explicit; rather, it expresses the fact that a certain essence belongs to a subject. For Aristotle the definiens presented in a definition expresses *what the subject is*, its essence. In chapter 2 we have seen that this is why Aristotle considers epistemic substances those things whose causes are the same as themselves. In contrast, the subject of a καθ' αὑτὸ συμβεβηκός predication is a cause of the existence of the predicate that is different from the predicate.[5]

Not only will the essence be predicated of a subject within a definition; the parts of the essence are not to be regarded as predicated of one another (90b35–38). Aristotle does not elaborate on why this is so, but, presumably, it is for the same reasons an essence is not to be thought of as "predicated of" that of which it is the essence. An essence is a single unity; the terms that denote parts of an essence do not denote different things.[6]

In considering whether one part of a definition can be predicated of another, Aristotle hints at a possibility that he will pursue in his actual solution to the problem. Perhaps it *is* possible that the same proposition can be made manifest by both a demonstration and a definition but *not* through having the definition itself be the conclusion of the demonstration in question. Rather, the definition would be revealed by showing why one part of the definition is predicated of another. Showing that such is in fact possible will be key to Aristotle's eventual solution to the problem facing him.

Aristotle also argues that a definition and a conclusion of a demonstration are to be distinguished semantically: an expression of what a thing is has a different meaning from an expression of the fact that a thing is something or other (ἔτι ἕτερον τὸ τί ἐστι καὶ ὅτι ἔστι[7] δεῖξαι [90b38; cf. 1.2.72a23–24]). Hence, if they are demonstrable, it can only

---

5. It is interesting to note that an implicit premise of this aporetic argument is that *no* demonstration has as its conclusion an implicitly definitional predication. This would mean that *every* demonstrable predication is καθ' αὑτὸ συμβεβηκός, making the problem of the logical structure of καθ' αὑτὰ συμβεβηκότα predications even more acute. The demand that demonstrable predication not be implicitly definitional, however, is not entailed by anything in Aristotle's account of demonstrations in *APo.* 1. Because the argument is aporetic, it could be interpreted as resting on premises Aristotle takes to be false.

6. Cf. *Metaph.* 7.12 and 8.6.1045a7–b7, in which Aristotle asserts that the unity of essence is not threatened by the multiplicity of the parts of the essence because every differentia implicitly contains and entails its genus. Aristotle's account in *APo.* 2.13 of the proper means of employing the method of division for securing a definition implicitly rests on this same teaching. See Barnes, *Aristotle's "Posterior Analytics,"* 234.

7. Here, as at 2.2.90a2, ὅτι ἔστι is employed to denote some state of affairs or fact.

be through different demonstrations, explaining different facts (2.3.90b38–91a2). Aristotle then denies that a single demonstration can simultaneously make manifest two different propositions, except for the special case in which the major term of one demonstration falls under the major term of the other demonstration, related "as a part to a whole" (i.e., as species to genus). For example, the demonstration that some attribute belongs to triangles can be used to demonstrate that this attribute belongs to all isosceles triangles. But an essence will not be the genus of the demonstrated attribute, nor will the genus be the essence of the demonstrated attribute (2.3.90b33–91a6).

The aporiai that Aristotle has presented indicate that, if there were a way to demonstrate a definition, the demonstration would somehow have to both explain the inherence of an attribute in a subject different from it and identify that attribute with some essence. We shall see that the series of syllogisms identified in 2.8 as a demonstration of an essence serves both of these functions.

B. *Posterior Analytics* 2.4

While the arguments of *Posterior Analytics* 2.3 against the demonstrability of a definition are based on general considerations regarding the roles demonstrations and definitions play within the sciences, 2.4 presents arguments to the same effect, which appeal to the syllogistic structure that such demonstrations would need to have.

First, Aristotle indicates that a subject term and the formula expressing the essence of that subject convert. To demonstrate that A is defined as C, more is involved than simply demonstrating that C belongs to A through a demonstration employing some middle term B. If A is the definiens of C, anything falling under A would fall under C, and vice versa.[8] So, if there were a demonstration that C is defined as A, it could not merely conclude that all A is C; it would also have to show that all C is A. (Of course, no Aristotelian syllogism can conclude both that all A is C and that all C is A, but we can

8. As D. Demoss and D. Devereux ("Essence, Existence, and Nominal Definition in Aristotle's *Posterior Analytics* II 8–10," *Phronesis* 33 [1988]: 149) point out, the thesis that all definitions convert is implicitly rejected in the resolution of the aporiai, for there Aristotle indicates that there will be one kind of definition, a nominal definition, which may give sufficient but not necessary conditions for the kind that is defined.

imagine a pair of syllogisms that share a middle term and, taken together, are considered a single demonstration, such as "All B is A, all C is B; all C is A" and "all B is C, all A is B; all A is C.") The middle term of such a "demonstration" would have to be coextensive with the extremes (2.4.91a14–18). But even this demonstration cannot be considered a demonstration of a definition, for at best it shows that C converts with A. We still need to demonstrate that C is the *essence* of C—in other words, that A belongs to C in what C is. *This* can be demonstrated only if A is in the essence of B and B is in the essence of C (91a18–25). (Here the transitivity of the relation "to be in the what it is" is assumed, but this is noncontroversial.) But this in turn requires a grasp of the essence of C, which is precisely what the demonstration was to ascertain (91a25–32).[9]

The rest of 2.4 is devoted to making Aristotle's point explicit through examples. One is Xenocrates' purported demonstration of the definition of soul:

> (i) All things that are the cause of their own life are self-moving numbers,
> (ii) all souls are the cause of their own life,
> so, (iii) all souls are self-moving numbers.

Even if Xenocrates' premises were to be granted, we would still have not demonstrated that "self-moving number" is the *essence* of soul. For, as shown, that would require additional premises in which the essence of soul is already assumed.

It is clear at this point that, if it were possible for a demonstration to show that a predicate has a certain essence, its logical form would have to be different from those of the demonstrations discussed in *Posterior Analytics* 1, in which the proposition to be made manifest is a conclusion of a first-figure universal syllogism whose premises are definitional predications. In 2.5 and 6 Aristotle considers other forms that a demonstration of a definition might take, only to reject them as well.

9. I generally follow the interpretation of the argument presented by Ross (*Aristotle's Prior and Posterior Analytics,"* 616–18). See also Barnes, *Aristotle's "Posterior Analytics,"* 199. Barnes, however, does not see that Aristotle is requiring that a demonstration of a definition somehow make clear that the definition of a subject is indeed its definition. Barnes' objections lose their force if this is kept in mind.

C. *Posterior Analytics* 2.5

In *Posterior Analytics* 2.5 Aristotle considers the method of revealing a definition by division, which Plato called an ἀπόδειξις.[10] The gist of Aristotle's argument (which covers some of the same ground as *Prior Analytics* 1.31) is that the successive determination of the subgenera into which a subject falls is not only nondeductive in form (2.5.91b12–17) but is also such that each step of the classification presupposes a deduction. If the definition at which one arrives is to be demonstrated, the division as a whole must be justified by a series of deductive inferences, each based on the definition of that whose essence is to be determined through the division.[11]

As an example, Aristotle considers how one might determine the essence of human beings (here taken to be "two-footed animal"). Let it be assumed that human beings are animals. Let it also be assumed that every animal is either footed or aquatic. (For this to be accommodated within Aristotelian syllogistic, "footed or aquatic" would have to be considered a single monadic predicate, but it is not clear that Aristotle realizes this. Nor is it clear whether Aristotle takes such a predication to be demonstrable; I doubt that he did.)[12] From this we can deduce that

10. See *Sophist* 242b4 and 258c10; *Statesman* 269c2 and 302e5.

11. This is why division is called "a feeble deduction" at *APr.* 1.31.46a33. See R. Smith, *Aristotle: "Prior Analytics"* (Indianapolis: Hackett, 1989), 160.

12. Barnes (*Aristotle's "Posterior Analytics,"* 201) holds that Aristotle does take such predications to be demonstrable and that they are what Aristotle has in mind at *APo.* 2.5.91b23–24: "To be sure, those who proceed in this way make nondeductive use even of what admits deduction." But this would mean that, according to Aristotle, the set of differentiae falling under a genus is deducible a priori on the basis of that genus. In that case one could deduce what all of the possible species are and then determine on the basis of empirical research which species are actually exemplified. It is true that at *Pol.* 4.4.1290b25–37 Aristotle writes as though this were the practice of the biologist, but it is highly unlikely that Aristotle's example here is meant to be taken seriously as a description of biological methodology. [Pellegrin ("Aristotle: A Zoology without Species," trans. A. Preuss, in *Aristotle on Nature and Living Things: Philosophical and Historical Studies Presented to David M. Balme on his Seventieth Birthday*, ed. A. Gotthelf [Pittsburgh: Mathesis Publications, 1985], 103–4)] thinks that the a priori procedure described here predates the empirical methodology of the biological writings. But then it is very odd that there is no trace of this procedure elsewhere. Indeed, this procedure could be considered a kind of demonstration of the definitions of the various biological species, the possibility of which is denied in *APo.* 2.3–7.) The genetic account of one's grasp of essences presented in *APo.* 2.19 indicates that Aristotle does not conceive of the scientist as deducing what possible essences there are and then hypothesizing the actual existence of those discovered to exist. Instead, he takes the scientist to simply hypothesize the reality of the essences of those simple kinds that are seen to exist. I confess, however, that it is unclear to me to what Aristotle refers in 2.5.91b23–24.

human beings are footed or aquatic. But on what grounds are we to deduce that human beings are footed, unless we already have a grasp of their nature (2.5.91b18–24)? This is why the definer assumes each term of the division and, hence, the whole conclusion.

Aristotle next indicates a problem with the method of division, which was well-known to Plato (*Phaedrus* 265e1–3): On what grounds is one sure that in dividing one is making the proper cuts in the genus? It is possible to proceed by division and reach a conjunction of predicates that converts with the subject yet does not express the essence of the subject (2.5.91b24–26). Further, without a prior grasp of the essence of the subject, what guarantee is there that the definition at which one arrives through division does indeed convert with the subject, delineating a class neither more nor less extensive than that of the subject (91b26–27)?

A division is possible only if one is already in possession of all of the definitional predications that, conjoined, constitute the essence of the subject. Nonetheless, even though a successful division must rest on the assumption of the definition, it does *display* the definition, making apparent the relations of class inclusion that the definitional predicates have to one another (2.5.91b28–35; cf. 2.13.96b28–97a6). For this reason, division can be considered an effective pedagogical device for presenting the results of the inquiry into the essence of a basic kind.

D. *Posterior Analytics* 2.6

Within *Posterior Analytics* 2.6 Aristotle considers two more ways a definition might be conceived as demonstrable. One might deem the following a possible demonstration:

> (i) A definition of A is the conjunction of all and only those attributes that are within the essence of A,
> (ii) B, C, and D are all and only those attributes that are within the essence of A,
> so, (iii) the definition of A is "a B that is C and D." (2.5.92a7–9)

Aristotle presents two objections. First, premise ii rests on a prior grasp of the essence (τί ἐστι) of A, which cannot be grasped without assuming the conclusion, the definition of A (92a9-10).[13] Second, a premise

13. On this argument, see Barnes, *Aristotle's "Posterior Analytics,"* 203.

such as i has no place within a true demonstration. If we are deducing that some attribute belongs to some subject A, any admissible premise would be of the form "B belongs to A" or "C belongs to A." Just as *syllogism* will never be the term of one of these predications, so the term *definition* must be absent from the premises.[14] Both what a deduction is and what a definition is are unstated assumptions that make possible the very discourse of the sciences (92a11–19).[15]

Aristotle's next aporia concerns another kind of proposed demonstration of a definition, presented in *Topics* 6.9.147a29–b25 and 7.3.153a26–b24, as a way in which one may dialectically argue for a definition. Aristotle shows how what may be an acceptable dialectical argument for a definition cannot be a demonstration. The proposed demonstration has the following form:

(i) A is defined as B,
(ii) Every term x is such that the contrary of x is defined as the contrary of the definition of x,
so, (iii) the contrary of A is defined as the contrary of B.

New definitions could thus be demonstrated by substituting the term denoting the contrary of x for "the contrary of x" and likewise with the definiens. For example, given that the bad is definable as the divisible, one could demonstrate that the good is the indivisible (2.6.92a20–24). Aristotle objects that the last step of this purported demonstration, in which predicates are substituted for predicates with the same reference, must rest on an implicit premise that the two terms involved in a substitution (that which is replaced and that by which it

14. I take this to be the import of 92a12: "the premise is always universal or particular." He means that a premise will always have as its subject a particular or universal member of the set of subjects that concern a deduction. Logical terms will not be members of this set (except, of course, in the special case in which the subject matter of that discourse is discourse, as in the *Analytics* itself.)

15. Perhaps those assumptions are among the common axioms mentioned at 1.2.72a16–18; 11.77a26–35. Ross (*Aristotle's "Prior and Posterior Analytics,"* 622–23) seems to be correct in taking Aristotle to be distinguishing certain logical principles from explicit *premises*. Barnes (*Aristotle's "Posterior Analytics,"* 203) takes issue with Ross on this, thinking that Aristotle is banishing all propositions regarding definitions and deductions from all reasoning. But Aristotle is not here concerned with all reasoning; indeed, *Top.* 7.3 makes clear that for Aristotle the sort of syllogism concerning definitions here being considered will have a place within a dialectical context. It is apparent from 2.5.92a6 that Aristotle wants to deny that this sort of deduction is a *demonstration*.

is replaced) are coreferential. It will be assumed that "the contrary of x" and some other term *y* are coextensive. On what grounds does one make this assumption? Is it to be taken as the conclusion of some prior demonstration? No, because a demonstrated conclusion, like any conclusion of an affirmative syllogism, takes the form of one term's being predicated of another. On the basis of such a conclusion alone, one would not know whether or not the predication converts (92a26–27).[16] Another possibility, which Aristotle does not consider, is that the identification of the reference of the two predicates is definitional. But then the substitution in question would rest on the definition of that subject whose definition was to have been ascertained through demonstration. It follows that the substitution cannot be employed in the demonstration proposed without begging the question.

All of the ways of demonstrating a definition Aristotle has considered and has dismissed were supposed to have revealed the definition in question by presenting it (in either the form "A is a B and C" or the form "the definition of A is 'B and C' ") as the conclusion of a syllogism. In *Posterior Analytics* 2.8–10 Aristotle discusses what he takes to be a demonstration that does in fact make a definition manifest. It will escape these criticisms because it will not present the definition in question as a syllogistic conclusion. Instead, it merely *enables* anyone comprehending the demonstration to formulate an appropriate definition of the predicate in question. In Aristotle's words the definition one arrives at will itself be like a demonstration but will "differ from a demonstration in how the terms fall" (2.10.94a2).

Aristotle concludes this set of aporiai by noting that all of the proposed demonstrations of definitions do not account for the unity of the definiens. If what is defined has an inner unity, this ought to be reflected in the definition. But all of the proposed demonstrations of definitions consider a definition to be made up of a set of conjoined predicates, with the order of that conjunction being irrelevant

16. One might object that whether or not a conclusion is convertible could be secured through two demonstrations. Aristotle does not address this objection. But he could on this point appeal to the theory of demonstration as put forward in *APo.* 1. For the minor term of one of these demonstrations must be the term to be defined, and Aristotle has made clear that any demonstration whose conclusion is a universal predication must ultimately rest on the definition of its minor term. Hence, such a demonstration of the definition must be regarded as question begging.

(92a27–33).[17] The metaphysical puzzle of how an essence that is expressible as a multiplicity of predicates can nonetheless have a true unity is not solved until the central books of the *Metaphysics*: a genus is each one of its differentiae in potentiality (*Metaph.* 7.12, 8.6). But the puzzle Aristotle is here raising is not this metaphysical one; rather, it is methodological and epistemological. How does the means by which we determine an essence make clear to us that what we are defining is some unity, the sort of thing that has an essence? This is not an issue Aristotle ever explicitly addresses.

Here it should suffice to indicate that Aristotle's genetic account in *Posterior Analytics* 2.19 of how one comes to be aware of the first principles lays a heavy emphasis on the soul's capacity to unify. Memories are unified into a unified experience (100a5–6), which in turn is responsible for the universal principle of knowledge (100b1–5). (The term for "universal," τὸ καθόλου, can be literally translated as "the in-respect-to-the-whole.") Presumably within this process, the soul is taken to be able to recognize something as having a true unity and to have its conceptual representation of the thing likewise unified, even though the essences of such unities are to be expressed as multiplicities. In contrast, the methods of determining a definition discussed in *Posterior Analytics* 2.5 and 2.6 consist of logical manipulations of a multiplicity of terms. Such reasoning cannot allow one to recognize a unity as such.

But the aporia is dialectical. For the things whose essences can, in a sense, be demonstrated will turn out not to have inner unities. These will be recurrent attributes, events, and phenomena, not the epistemic substances.

E. *Posterior Analytics* 2.7

The aporetic arguments against the demonstrability of definitions that we have thus far examined are sound. Their conclusions logically follow from premises that would be accepted by anyone who accepts the account of scientific explanation that Aristotle has thus far presented. More specifically, they rest on the understanding of a definition as a kind of universal, convertible proposition that serves as a first principle

17. As Ross (*Aristotle's "Prior and Posterior Analytics,"* 624) points out, *Top.* 7.3.153a23–b24 makes clear that this will be a concern for one arguing for a definition based on the basis of considering the definition of the contrary of what is to be defined.

and on the assumption that a demonstration of a proposition is a first-figure syllogism with that proposition as its conclusion. The aporiai are to be solved by recognizing that there are other kinds of definitions. For this reason, they are to be solved by adding to the account of scientific explanation already presented, not by refuting its basic assumptions. In contrast, all three aporetic objections against the demonstrability of definitions presented in 2.7 would be recognized as unsound by one familiar with the theory of demonstration presented in book 1. But, even though Aristotle's arguments here are faulty, they reveal conceptual difficulties whose resolution leads to the correct way of viewing the issue at hand.

After reviewing the problem as presented in the previous chapters, Aristotle asks on what grounds one who purportedly demonstrates a definition knows that the subject being defined exists. The passage can be translated as follows:

> Another point is this. How will one who is defining prove the "what it is"? For one who knows what a human being or anything else is must also know that it exists (ὅτι ἔστι)—for no one knows what a nonexistent thing is (τὸ γὰρ μὴ ὂν οὐδεὶς οἶδεν ὅ τι ἐστιν) but, while you know what the phrase or word means (τί μὲν σημαίνει ὁ λόγος ἢ τὸ ὄνομα) when I say "goatstag," it is impossible to know what a goatstag is (τί δ' ἐστὶ τραγέλαφος ἀδύνατον εἰδέναι). But if you are demonstrating what it is *and* that it exists, how are you going to do this by the same argument? For definition makes one thing clear, and so does demonstration. But what a human being is and the fact that human beings exist are different (τὸ δὲ τί ἐστιν ἄνθρωπος καὶ τὸ εἶναι ἄνθρωπον ἄλλο).
>
> Next, we say that it is necessary to demonstrate every nonessential attribute of a thing.[18] But existence (τὸ εἶναι) is the

18. I follow the reading of Ross: Εἶτα καὶ δι' ἀποδείξεώς φαμεν ἀναγκαῖον εἶναι δείκνυσθαι ἅπαν ὅ τι ἐστίν, εἰ μὴ οὐσία εἴη. A literal translation of the passage is: "Next we say that it is necessary to show that something is through demonstration, unless it is the essence." On this reading Aristotle claims that every attribute of a subject is demonstrated to hold of a subject except for the whole essence of that subject. (See ibid., 626–27.) But there are also unmediated predications, each of which is a part of a whole definition expressing the whole essence. These unmediated simple predications, too, will not be demonstrated. Aristotle is writing sloppily, as he sometimes does. By "unless it is an essence" I take Aristotle to mean "unless it is an essential attribute."

> essence of no subject, since existing (τὸ ὂν) is not a genus. Therefore, there will be a demonstration of a subject's existence. And this is the very thing now done by the sciences. For the geometer assumes the meaning of "triangle," but he demonstrates that there is such a thing (ὅτι ἔστι). So of what will the one who defines "triangle" prove the existence other than triangle?[19] Then it will come about that he does not know whether it exists, even though he knows by definition what it is. But that is impossible. (2.7.92b4–18)

Aristotle's abbreviated argument here can be reconstructed as follows. He begins by appealing to the requirement of 2.1–2 that an inquiry into τί ἐστι (what a thing is) either accompany or be preceded by a positive resolution of the inquiry into εἰ ἔστι (whether a thing exists). Within the exposition of a science, the existence of a subject must be presented prior to its essence. Otherwise, one would know what something is without knowing of its existence, which is impossible (2.7.92b4–5, 17–18).[20]

Then, for the first time within the *Posterior Analytics*, Aristotle explicitly recognizes that there is a kind of definition that is such that it can be possessed by one who does not know whether the referent of the definiendum exists. One may have such a definition of something that is indeed nonexistent. This is an account of what a term means (τί σημαίνει), as contrasted with an account of what a thing is (τί ἐστι).[21]

The manuscripts read ὅτι ἔστιν for ὅ τι ἐστίν. We could then translate 92b12–13 as: "Next we say that it is necessary to demonstrate any assertion of existence, unless the subject is an οὐσία." (Cf. J. Tricot, *L'Organon d'Aristote, traduction nouvelle et notes*, 2d ed. [Paris: J. Vrin, 1979], 185.) This is an attractive alternative; it would offer support to my thesis that the subjects whose existence is not demonstrated are beings the sciences consider substantial. But it does not give Aristotle a cogent argument in what follows.

19. The text is unintelligible as it stands. I here follow the emendation of the text suggested to me by Sarah Broadie: ὅτι ἔστιν ἢ τὸ τρίγωνον for τί ἐστιν ἢ τρίγωνον, at 92b17.

20. Although 2.7.92b17–18 come at the end of the argument we are considering, they serve as a reprise of its crucial premise. The lines ought to be read as a response to one who denies the conclusion of the argument; doing that entails denying the premise, which is absurd.

21. Aristotle does not always keep the terminological distinction that he is making here. Thus, at 1.1.71a15–16 and 1.10.76a32 he uses τί σημαίνει to refer to a scientific definition of an epistemic substance (though at 1.10.76b6–7 he seems to distinguish between the definition assumed of an epistemic substance and that assumed of a derivative entity by calling the first an assumption of τοδὶ εἶναι and the latter an assumption of τί σημαίνει).

The first, which I, following tradition, call a nominal definition,[22] is the account of the sense of a term, the way in which one goes about determining the reference of a term (if it does indeed refer). The second is an expression of an essence, which can serve as a principle of scientific explanation. The fact that we would know when to call a beast "a goatstag," even though there are no beasts that we would actually call by this name, confirms that there can be nominal definitions of nonexistent kinds, though these cannot be employed as demonstrative premises. In the present passage the distinction is made with the ostensible aim of focusing attention away from nominal definitions.

Aristotle remarks that it is possible to comprehend a nominal definition without knowing the existence of the definiendum. This is because the truth of this definition is dependent on linguistic usage, not on the things to which words refer. But the possibility of giving nominal definitions for nonreferring terms is irrelevant to the problem at hand. For we are concerned with whether *scientific* definitions can be demonstrated, and for *these* one must already be assured of the existence of what is being defined. We shall see, however, that nominal definitions will indeed have a large role to play in the "demonstration of a definition" discussed in 2.8.

Since a scientific definition presupposes the existence of the definiendum, such a definition presupposes an account by which this existence is posited or justified. A demonstration of a definition, if there were such a thing, would therefore have to either express both this essence and its existence concurrently, or rest on some other expression, which will justify the existence of the definiendum. Aristotle begins by eliminating the first possibility. He points out that the way in which we express the essence and existence of human beings (τί ἐστιν ἄνθρωπος and τὸ εἶναι ἄνθρωπον) are different. Accordingly, the definition and the expression by which existence is scientifically ascertained will not be one and the same expression. Could the demonstration of the definition rest on some other expression, by which the existence of the definiendum is made an object of scientific understanding?

Aristotle asserts that only definitional attributes of a thing (those that constitute its οὐσία) are assumed and denies that being (here having the sense of "existence") is the genus of anything, leaving unstated

22. See chap. 3, sec. E.

the obvious truth that being cannot be a differentia either.[23] Hence, the existence of the subject is not assumed.

It appears as though the only remaining possibility for a demonstration of a definition would be an expression having the form of being a demonstration of *both* the existence and essence of the subject in question. But, since the identification of the subject's essence and existence are two different propositions (cf. 1.2.72a23–24), the explanatory account of its essence cannot be the explanatory account of its existence. The conclusion of the demonstration in question cannot express both of these (92b8–11). If the existence of the subject is to be the object of a science at all, it is to be demonstrated, just as geometers demonstrate the existence of triangles (by showing how they can be constructed) (92b12–16).[24]

Both text and argument at 92b17–18 are exceedingly obscure. I suggest the following interpretation. Aristotle here shows that it cannot be the case that the demonstration of the existence of one subject acts as a demonstration of the existence of another subject; one who defines "triangle" will need a concurrent or prior demonstration of the existence of just that: triangle (2.7.92b17).[25] This eliminates one last manner in which one may think that a subject's definition can be demonstrated.

The careful reader of the first book of the *Posterior Analytics* will balk at Aristotle's assertion that the existence of a kind is to be demonstrated. For, as we have seen (chap. 3, sec. D), Aristotle had there explicitly distinguished those for which this is to be assumed. Aristotle has already indicated that it is not *in every case* true that the expression of a thing's essence is to be made clear by a definitional first principle, as is

---

23. *Top.* 4.1.121a18–19, b16; and *Metaph.* 7.16.1040b16–24 and 8.6.1045b1–7, are Aristotelian passages denying that being is a differentia. (See Barnes, *Aristotle's "Posterior Analytics,"* 205–6.) Aristotle presents a dialectical argument for being's not being a genus at *Metaph.* 3.3.998b22–27. This argument rests on the questionable premise that a genus is properly predicated of that of which the differentia is predicated, not the differentia itself. While, for example, "animal" is properly predicated of "human" but not of "rational," "existent" *is* predicated of both "human" and "rational." It is arguable that within *Metaph.* Aristotle does come to regard being as an all-encompassing generic kind: "what is insofar as it is," studied by the science called first philosophy (cf. *Metaph.* 4.1.1003a21–22). But *APo.* 1.32 makes clear that Aristotle here denies that there are any principles applying to all genera. Accordingly, Aristotle did not at the time of *APo.* consider being a genus, with its own καθ' αὑτά attributes, for every such kind is the proper concern of a single science.

24. See chap. 3, sec. D.

25. See n. 19.

assumed in the aporetic argument of 2.7. For the definitions of the entities studied by a science that are not epistemic substances will not have the status of being first principles of demonstration. For these beings, the science assumes only the statement of the meaning of the terms denoting them. Further, Aristotle has already indicated that it is not true *in every case* that the existence of a kind is demonstrated. The existence of an epistemic substance is assumed by the science studying it, for such a science does not consider the epistemic substance as inherent in any other entity.

Aristotle's aporia raises true problems only in the case of those kinds that are not epistemically substantial, including the καθ' αὑτὰ συμβεβηκότα. For these *are* demonstrated to exist. Could it be the case that the demonstration of the existence of such things can somehow serve as a definition, or is this impossible, as Aristotle indicates in the aporia of 2.7.92b17–18?

Aristotle considers the sort of demonstration that explains the existence of a καθ' αὑτὸ συμβεβηκός attribute in 2.8. As we shall see, one of the premises of Aristotle's aporetic argument here—that an explanatory account of existence cannot at the same time be an explanatory account of essence—does not hold in regard to this sort of demonstration. The demonstration explaining why a subject is to be identified as having a certain essence will turn out to be the same as a demonstration of that subject's existence.

In *Posterior Analytics* 2.7 Aristotle presents two additional dialectical passages bearing on the issue of the demonstrability of demonstrations. Both of these, like the first, rest on questionable premises, the consideration of which points to the proper resolution of the problem. In the second passage Aristotle does not so much present an objection against purported demonstrations of definitions as raise an objection against the form of definitions themselves. Not only does the assertion that a circle is a figure all of whose points are equidistant from its center not make clear that such an entity does in fact exist;[26] it does not make clear

26. The passages at 92b19–21 and 23–25 do not constitute an independent argument. Rather, through an appeal to contemporary definitions familiar to his audience, Aristotle confirms that definitions do not explain the existence (or even the possible existence) of the definienda. We have seen that this is a premise of the previous argument, to the effect that the expression of an explanation of a thing's existence will be different from the definition of that thing, which expresses its essence. However, the point Aristotle makes at 92b21–22 is new and independent.

that this account describes those figures we recognize as circular (2.7.92b19–25). An adequate definition must somehow display the cause of its being the definition of what is being defined.

As it stands, this objection is easily countered on the basis of the theory of demonstration presented in *Posterior Analytics* 1. Definitions are first principles, characterized by their not demanding explanation.[27]

Again, the aporia is not without value for the solution of the problem. It indicates a demand that would have to be met by a demonstration of a definition, were there to be such a thing. Any definition that would be such as can be rendered intelligible by a demonstration would require some kind of justification. But exactly what would this involve? A nominal definition as such needs no justification outside of an appeal to the linguistic usage (of either the definer alone or some linguistic community). But suppose we had a scientific explanation of the existence of a certain thing, for which there is a nominal definition of the term denoting it. One would require a justification of the identification of the referent of the term that is defined with that entity whose existence is explained by the definition. Employing Aristotle's example, suppose that the existence of circles is to be demonstrated geometrically and is not assumed.[28] Circles would therefore not be among the epistemic substances. Accordingly, the definition of a circle would not be a demonstrative first principle; instead, it would be a nominal definition. If we take seriously Aristotle's point in this passage, we are led to wonder how one in possession of the demonstration of the existence of circles will be assured that the thing whose existence has been demonstrated is indeed a circle (taking *circle* to denote whatever satisfies the nominal definition of the term).

When Aristotle does give an account of demonstrations of definitions, he addresses this very point. A demonstration of the existence of a καθ' αὑτὸ συμβεβηκός entity proceeds by explaining why that which is described by a nominal definition of a term in fact exists. The conjunction of the cause of this entity and those of its features by which we recognize it (which will be those features expressed in the nominal definition of the term denoting that entity) will serve as the scientific definition of the entity. This is why the demonstration by which the exis-

27. See J. Hintikka, "On the Ingredients of an Aristotelian Science," *Nous* 6 (1972): 59; A.P.D. Mourelatos, "Aristotle's 'Powers' and Modern Empiricism," *Ratio* 9 (1967): 102.

28. The existence of circles *is* assumed in postulate 3 of Euclid's *Elements*, bk. 1.

tence of that entity is explained will also serve as a justification of the scientific definition of that same entity.

Having purportedly shown that there can be no demonstration of a thing's essence, Aristotle closes this aporetic section of the *Posterior Analytics* by briefly considering another way in which there could be demonstrations of definitions: perhaps the demonstrator shows that a certain *term* has a certain meaning (2.7.92b26–28). The conclusion of such a demonstration would be a nominal definition, which, since it has been demonstrated, would qualify as an object of science and, accordingly, could itself be employed to generate premises of demonstration. Aristotle does not find it necessary to argue at great length against this suggestion. There can be nominal definitions of things that do not exist.[29] Further, one can, through a nominal definition, stipulate that a term will signify any concatenation of predicates, whether those predicates constitute a true unified essence or not. A description of Homer's first epic could be considered a definition of the term *Iliad* (2.7.92b28–34), even though such a description need not have any explanatory role or express that which makes Homer's poem a unity.[30] None of this is acceptable because, while nominal definitions are statements concerning words and concepts, the sciences are concerned with *things*. Scientific definitions must express things as they really are. Nonetheless, as we shall see in the next chapter, a nominal definition will in a sense be the conclusion of the demonstration of an essence that Aristotle endorses.

We have seen that in *Posterior Analytics* 2.3–6 Aristotle presents solid

29. From the fact that one can signify nonexistents (92b29), it follows that there can be expressions that fix the signification of those terms that do signify nonexistents. Hence, Aristotle is presenting an argument in favor of there being nominal definitions of nonexistents. On the other hand, Aristotle wants to resist saying that there can be definitions of nonexistents, on the grounds that definitions, being objects of scientific knowledge, can only express the nature of things as they are. The aporia arises because, in the present context, Aristotle has not distinguished between those definitions that fix the signification of a term (nominal definitions) and those that have the status of objects of sciences. The aporia is resolved by making this distinction, as is done at 2.10.93b38–39. By not recognizing the aporetic character of the argument, Bolton ("Essentialism and Semantic Theory in Aristotle's *Posterior Analytics* II.7–10," *Philosophical Review* 85 [1976]: 521, 527–28), Sorabji ("Definitions: Why Necessary and in What Way," in *Aristotle on Science: Proceedings of the Eighth Symposium Aristotelicum*, ed. E. Berti [Padua: Editrice Antenore, 1981], 218), and Demoss and Devereux ("Essence," 140) mistakenly take Aristotle to be denying that there can be nominal definitions of nonexistents.

30. Cf. *Metaph.* 7.4.1030a7–11, 1030b8–11, 8.6.1045a12–14; *Poet.* 20.1457a28–30.

arguments against the possibility of there being demonstrations of definitions. The subjects considered by a science have their definitions assumed within that science, and any "demonstration" having such a definition as a conclusion would have to be at least implicitly circular. These objections hold only if all definitions serve as first principles, but, as we have seen, the definitions of καθ' αὑτὰ συμβεβηκότα predicates will not serve as first principles. In 2.7 Aristotle presents arguments concerning the logical and semantic relations that hold among expressions and explanations of essence and existence. Although these arguments rest on ignoring some important distinctions that Aristotle had already made, they serve to draw our attention to those entities that must be demonstrated to exist by a certain science: the καθ' αὑτὰ συμβεβηκότα. Although these entities may have a place within the exposition of a science, they will not be among the subjects defined by a science, nor will they be predicated in the definitions of the subjects that are so defined. This is why there is a real theoretical problem concerning how the existence of such predicates will be demonstrated. As we shall see, by solving this problem, one can show how it is the case that the demonstration of a καθ' αὑτὸ συμβεβηκός predication serves to reveal and thereby justify the scientific definition of the predicate.

5

Inquiry and Explanation: *Posterior Analytics* 2.8

Posterior Analytics 2.8–10 have given rise to two main lines of interpretation. One is derived from the commentary of Philoponus, the other from the paraphrase of Themistius. The interpretation that I will be offering is, like that of Ross and Landor, along the lines of Themistius' understanding of the text. But within both the tradition and recent Aristotelian scholarship, interpretations along the lines of Philoponus' commentary have predominated.[1]

It is not unusual for a text of ancient philosophy to give rise to divergent interpretations. What is unusual about the case of *Posterior Analytics* 2.8–10 is that the text at issue is long, and the points at issue are basic: What is Aristotle's subject, and what is he saying about it? To this question each interpretation offers radically different answers. Accepting the interpretation of Philoponus, or one like it, involves rejecting the main thesis of this book—that within 2.8 Aristotle presents the logical form of an explanatory demonstration of a καθ' αὑτὸ συμβεβηκός predication.

For this reason, I begin my account of *Posterior Analytics* 2.8–10 by indicating what I take to be the deficiencies of the prevailing line of interpretation of these chapters. There are, of course, disadvantages to this approach. I shall have to discuss passages out of context, with an eye to why certain views about them are wrong, instead of considering what the correct view is. But, by proceeding in this way, I address probable objections at the start, better allowing my interpretation to be evaluated on its own terms.

I consider two representative interpretations that take similar

1. McKirahan (*Principles and Proofs: Aristotle's Theory of Demonstrative Science* [Princeton: Princeton University Press, 1992], 297 n. 1) is a notable exception. He dismisses Mansion's line of interpretation merely by indicating that he finds it without textual or philosophical support.

approaches, the ancient interpretation offered by Philoponus and the more recent treatment by Suzanne Mansion.

A. Philoponus on *Posterior Analytics* 2.8

I begin by sketching Philoponus' reading of *Posterior Analytics* 2.8. Aristotle begins this text by remarking that, having presented aporetic arguments against the possibility of there being a demonstration of an essence, the whole issue must be reconsidered (93a1–3). Aristotle again asserts that it is the same thing to know the cause of something and the τί ἐστι (the "what is it," or essence) of that something (93a3–4; cf. 2.2.90a14–15). But, Aristotle notes, there is more than one kind of identification of a cause.

> The account of this is that there is some cause, and this is either the same or different, and if it is different, either demonstrable or indemonstrable. If now it is different and it is possible to demonstrate (ἐνδέχεται ἀποδεῖξαι), it is necessary that the cause be a middle term and be shown (δείκνυσθαι) in the first figure. For what is shown is universal and affirmative. (93a5–9)

It is no surprise that a passage as unclear as this should give rise to important differences of interpretation. What is being distinguished? On the interpretation for which I shall be arguing, Aristotle, consonant with what precedes, identifies the "what is it" of any thing or kind as a single, privileged cause of that thing or kind. When Aristotle asserts that there are both demonstrable causes and indemonstrable causes, he is making a distinction between two different kinds of things: those whose causes are (in some sense) "demonstrable" and those whose causes are not. Philoponus, in contrast, takes the distinction here between cases in which the cause is the same and those in which the cause is different not as between two different kinds of subjects (distinguished by their having two different kinds of cause) but, rather, as between two different accounts that express a cause of the same thing. As an example, Philoponus appeals to the partial definitions presented in *De Anima* 1.1.403a25–b16. A material definition of anger, that is, one that is "taken from what is present in the matter" (λαμβενόμενος ἐκ τῶν προσόντων τῇ ὕλῃ), is "a boiling of the blood around the heart," while the formal definition, that is, one that is taken from what is present in the essence of the thing (λαμβενόμενος ἐκ τῶν οὐσιωδῶς ὑπαρχόντων τῷ πράγματι) is "a desire for returning pain for pain."[2] Because neither definiens converts with its definiendum, neither account of anger expresses a cause of anger that is "the same as" anger itself. One with a fever may have the boiling of the blood characteristic of anger without being angry, and, likewise, a cool-headed dissimulator may plot pain in return for pain without having his or her blood boil, a condition for true anger. Only the whole definition, expressing both the material and formal aspects of a thing's essence, can be said to express a cause that is the same as the thing itself.

In the following lines, 2.8.93a9–13, Aristotle reiterates that a demonstration will be a first-figure universal syllogism whose middle term expresses the "what it is" (or essence) of the minor term.[3] Philoponus sees this as an account of how one might "demonstrate" a material definition of x by taking the formal definition of x as a middle term. For example, the definition of anger would be "demonstrated" as follows:

(i) Anger is a desire for revenge,
(ii) everyone who desires revenge has a boiling of the blood around the heart,
so, (iii) anger is a boiling of the blood around the heart.

But, as noted, Philoponus sees that premise ii of this syllogism is not true; "boiling of the blood around the heart" is predicated neither universally nor necessarily of "desire for revenge." The major premise is merely ἔνδοξος, or generally admitted.[4] Philoponus thus concludes that this syllogism is not truly demonstrative, and has a role only in a dialectical context. This is taken to be the meaning of 93a14–15: "Now it was said earlier that this way is not a demonstration, but is a dialectical (λογικός) syllogism of the 'what it is.'"[5]

2. J. Philoponus, *In Aristotelis Analytica Posteriora Commentaria*, ed. M. Wallies, *Commentaria in Aristotelem Graeca* 3, no. 13, Prussian Academy Edition (Berlin: George Reimer, 1909), 364, 24–28.

3. For a translation of these lines, see sec. C.

4. This answers Landor ("Aristotle on Demonstrating Essence," *Apeiron* 19 [1985]: 121): "It is unclear why Philoponus thought the premises of such syllogisms are not necessary."

5. Landor ("Aristotle," 117) argues that Philoponus' line of interpretation is questionable at the outset because Aristotle has just reiterated that the middle term of a demonstration explains *the existence of something*, while for Philoponus that which the middle term explains is *a partial definition*. But Philoponus can adequately meet this objection.

Under Philoponus' interpretation Aristotle nonetheless proceeds to discuss this dialectical syllogism. Aristotle is thought to show how it is possible for us to employ such a syllogism to confirm the "material definition," not how it is possible for such a syllogism to function as a truly explanatory demonstration (for the latter is not possible). It is in this light that Philoponus interprets the rest of 2.8, up to its conclusion (93a16–b14). The account that Aristotle here presents of the various kinds of initial knowledge of a thing's existence, and how they are related to the endeavor to ascertain its essence, is both obscure and important, but the details of Philoponus' interpretation of this need not now concern us. It is enough to note that Philoponus considers Aristotle to be discussing the deduction of a material definition through a formal definition throughout. For example, the "demonstration" of the essence of thunder sketched at 93b8–12 is thought to be:

(i) Thunder is a quenching of fire,
(ii) quenching of fire is a noise in the cloud,
so, (iii) thunder is a noise in the cloud.

Here "noise in the cloud" is considered by Philoponus the material definition of thunder, and "quenching of fire" its formal definition.[6]

There are two major weaknesses to Philoponus' interpretation. First, it is not clear how "noise in the cloud" is taken from the "matter" of thunder any more than is "quenching of fire." Perhaps by "matter" here Philoponus simply means "substrate" or "subject" (ὑποκείμενον), but, even so, it is not clear that the cloud is the real substrate of the thunder rather than merely what is denoted by the grammatical subject of which "thunder" is predicated. After all, one would expect that a noise would inhere in the substance (water or fire) that is producing it, rather than the thing within which it is to be found (the cloud). This issue demands fuller discussion, but Philoponus does not provide it.[7]

The partial definition implicitly expresses the existence of the definiens, for example, noise in the clouds. For any true predication entails the existence of its subject and, hence, of the predicate. Cf. J. Barnes (*Aristotle's "Posterior Analytics," Translated with Notes* [Oxford: Clarendon Press, 1975], 208).

6. Philoponus, *In Aristotelis*, 369, l. 30–370, l. 15; 373, l. 26–374, l. 24.

7. One possibility is that "noise," as the genus of thunder, serves as a kind of "matter," to be further determined by the differentia of thunder. This is the strategy of Mansion, discussed in sec. B.

There is a second and more significant problem. We have seen that Philoponus' interpretation rests on taking 2.8.93a5–7 ("The account of this is that there is some cause, and this is either the same or different, and if it is different, either demonstrable or indemonstrable" [λόγος δὲ τούτου, ὅτι ἔστι τι τὸ αἴτιον, καὶ τοῦτο ἢ τὸ αὐτὸ ἢ ἄλλο, κἂν ᾖ ἄλλο, ἢ ἀποδεικτὸν ἢ ἀναπόδεικτον]) as making a distinction not between different kinds of subjects but, rather, between different kinds of explanatory accounts. The Greek does admit this reading. But the same cannot be said of two other passages in which Aristotle makes similar distinctions: 2.8.93b18–19 ("so that in the case of something of which there is a different cause it is not possible to know that 'what it is' without demonstration" [ὥστ' οὔτ' ἄνευ ἀποδείξεως ἔστι γνῶναι τὸ τί ἐστιν, οὗ ἔστιν ἄιτιον ἄλλο]) and 2.9.93b21 ("of some things something else is the cause, but of others it is not" [ἔστι δὲ τῶν μὲν ἕτερόν τι αἴτιον, τῶν δ' οὐκ ἔστιν]). In these passages Aristotle is clearly making the point that some subjects, as opposed to other subjects, have a cause that is "different" and that only in these cases is a demonstration of the essence possible. Philoponus does not present any account of the phrase οὗ ἔστιν ἄιτιον ἄλλο at 93b19 and interprets the distinction at 93b21 as being between those things with a cause that is temporally prior to them and those without such a cause. But this is highly unlikely, since all three distinctions are made in almost identical terms.[8] It is necessary to provide an account of 2.8 and 2.9 that takes seriously Aristotle's assertion

8. Bolton ("Definition and Scientific Method in Aristotle's *Posterior Analytics* and *Generation of Animals*," in *Philosophical Issues in Aristotle's Biology*, ed. A. Gotthelf and J. Lennox [Cambridge: Cambridge University Press, 1987], 130–42) presents an interpretation of this passage much like that of Philoponus (to whom the debt is acknowledged on 140 n. 35) and, consequently, runs into similar problems. Taking an interpretation of the distinction along the lines of Ross and myself to be "standard," he writes, "it is important to see that Aristotle does not focus his *main* attention there on two types of entities distinguished in the way supposed by the standard interpretation described above. The main items distinguished in this way are rather two different types of 'cases of what a thing is.' Nothing which he says shows that these accounts cannot be different types of cases of what the *same* thing is" (142). The Greek will not support Bolton's point. Bolton would take the referents of τῶν at 2.9.93b21 and 25 to be not the things to be explained but, rather, the causal accounts of those things. 93b25–26 would then have Aristotle picking out some *causal accounts* for which *their* essences (and not the things of which they are accounts) have a middle term and a different cause. But that is strained and obscure, even for Aristotle. This point is made in regard to the way Philoponus would have to read this passage by Landor ("Aristotle," 119–20).

that the sort of explanation he has in mind is possible only in respect to those things that, in some sense, have "no different cause."[9]

B. Mansion on *Posterior Analytics* 2.8

Philoponus' line of interpretation was enormously influential in the Aristotelian tradition.[10] A recent and prominent representative interpretation along Philoponus' lines is that of Suzanne Mansion, put forward in *Le jugement d'existence chez Aristote.*[11]

Like Philoponus, Mansion takes Aristotle to be showing how the material definition of a thing can be "demonstrated" through its cause, the "formal definition." But Mansion recognizes that the Greek text at 93a5–7 and 93b18–21 must be interpreted to the effect that the demonstration Aristotle is sketching is only applicable to certain *kinds* of things: those whose causes are different from themselves. She argues that Aristotle makes this distinction because the syllogism he sketches is possible only in the case of those things that have a material definition different from their formal definition (or their "cause").[12] But what sort of being will not have a material definition different from its formal definition? Mansion recognizes the temptation to say that all substances are of this class: if a definition of a substance is an expression of its form, one who apprehends the genus and differentia of a substance will be in possession of the whole definition. But Mansion resists this move. She appeals to *Metaphysics* 7.17.1041a26–30, b5–9, which she argues serves to show how one can explain what it is to be a human substance by showing that a certain form, the rational soul, expressed

9. There are other problems with Philoponus' interpretation of what has "a different cause." According to Philoponus (*In Aristotelis*, 371, 4–27), in 93a16–29 Aristotle focuses on a cause that is different from the thing on account of its being temporally prior to that thing (and hence is its efficient cause). [Cf. the interpretation of Aquinas (*In Aristotelis Libros Peri Hermineias et Posteriorum Analyticorum Expositio*, ed. R.M. Spiazzi [Turin, It.: Marietti Editori, 1955], L.II,l.vii,477)], who explains that, if the cause of a thing is the same as the thing, it is the thing's essence, that is, its form or matter, while, if the cause of a thing is different from the thing, that cause will be the thing's final or efficient cause.) But in these lines Aristotle is not concerned with the temporal sequence of the thing and its cause; he is, rather, concerned with the temporal sequence of our *awareness* of the two.

10. See, for example, Aquinas, *In Aristotelis*, L.I,l.vii,481; and J. Pacius, *In Porphyrii Isagogen et Aristotelis Organum, Commentarius Analyticus* (Hildesheim, Ger.: Olms, 1966), 326.

11. S. Mansion, *Le jugement d'existence chez Aristote*, 2d ed. (Louvain: Publications Universitaires de Louvain, 1976), 183–97.

12. Ibid., 186.

in the formal definition of "man," is present in matter having certain characteristics, the account of which is the material definition of man.[13] By so revealing the reason for the existence of a certain composite substance, a human being, one is showing what it is to be such a substance: a certain form within certain matter. She identifies "that of which there is no different cause" with the "simples" of *Metaphysics* 7.17.1041b9–11. What these are she leaves unclear.[14]

Like Philoponus, Mansion appeals to Aristotle's distinction between the material and formal definitions of anger presented in *De Anima* 1.1 as an example of the distinction made in *Posterior Analytics* 2.8. Unlike Philoponus, she recognizes that, while "anger is a boiling of the blood around the heart" can be considered a material definition in the sense in which it expresses the material cause of the subject, this cannot be the case for "thunder is a noise in the clouds." Rather, the two definitions are both "material" in the sense in which they express the *external characteristics* of what is defined. Similarly, "anger is a desire for revenge" and "thunder is a quenching of fire" are both formal insofar as they determine the material definitions and render them precise.[15]

But these two senses of "material definition" are so different that it is impossible to read Aristotle as presenting a single argument to the effect that the essence of all composites is demonstrated on the basis of their material definitions. "A human being is flesh and blood" is material in the sense in which it expresses the material cause of a particular composite substance; "an eclipse is a blocking of light" is material in the sense in which it expresses the full definition in an indeterminate manner. According to Mansion, one who demonstrates a material definition of "human being" in the way described in *Posterior Analytics* 2.8 will not show how the form of being human inheres in a material substrate (as described in *Metaph.* 7.17) but will present a certain syllogism whose conclusion is some subset of the predications that constitute the essence of man (as essence is understood in the *Posterior Analytics*, in which, because the notion of matter is absent, the essence of a thing is not taken to express its matter). For example, the demonstration of the

13. Ibid., 195–96.

14. Mansion refers the reader to a discussion by O. Hamelin (*Le système d'Aristote* [Paris: F. Alcan, 1920], 114–16). In the text cited Hamelin takes these simples to encompass two groups: the highest genera, or categories, and immaterial substances, such as the unmoved movers.

15. Mansion, *Le jugement*, 188, 196.

essence of man would have as its conclusion "man is an animal" (cf. 2.8.93a23–24). But such predications are unmediated.[16]

I suggest that endeavors to follow the tradition and see Aristotle as describing how one can demonstrate a material definition on the basis of a formal definition are unfruitful, unless *material definition* here simply means "part of a definition," and by *formal definition* one means what Aristotle in the text calls a "cause." Since the appeal to Aristotle's later form-matter distinction does not clarify much, I will not employ it in my own exposition and discussion of Aristotle's text. Because the attempt to see Aristotle as showing how one part of an indemonstrable definition can explain another leads one into difficulties, it is worth considering the merits of an interpretation according to which Aristotle is showing how what one thing is can be explained on the basis of the definitions of other things. I now present such an interpretation.

C. Preliminary Apprehensions of What Is to Be Explained

Aristotle begins the argument of *Posterior Analytics* 2.8 as follows:

> Since, as we said, it is the same thing to know what something is and to know the cause of its existence (τὸ αἴτιον τοῦ εἰ ἔστι)—and this is the reason for this (λόγος δὲ τούτου): there is a cause, and the cause is either the same or different, and if the cause is different, it is either demonstrable or indemonstrable—if the cause of something's existence is different and this admits demonstration,

16. These difficulties with Mansion's thesis were first presented by C. Rutten ("Note sur le syllogisme de l'essence chez Aristote," *Phronesis* 9 [1964]). Rutten agrees with Mansion that within the *APo.* Aristotle envisages a "demonstration of the essence" in the case of substances but, unlike Mansion, takes Aristotle to be "promising the impossible." He argues that, because the explanation of the being of a composite substance that is presented in *Metaph.* 7.17 presupposes the essence of "human being," it is evidence that at the time of the *Metaph.* Aristotle recognized that there could be no "demonstration" of the essence of a substance.

As evidence of the claim that Aristotle posited "demonstrations" of the essence of substances in the *APo.*, Rutten adduces 2.2.90a9–11, in which Aristotle appears to assert that the cause of the being of substances is expressed by a middle term. But the examples of the Moon, Earth, and the Sun presented at 90a12–13 can be considered as examples of *subjects*, the sort of thing that can be ἁπλῶς (in an unrestricted sense). Aristotle need not be read as making the claim that there is a middle term making possible an explanation of the existence of substances, the ultimate metaphysical subjects.

> the middle term must be the cause and the demonstration must be in the first figure. For that which is demonstrated is universal and affirmative. (93a3–9)

This passage is as important as it is obscure. Aristotle begins by returning to his identification of the "what it is" and the "why it is" at 2.2.90a14–15. We recall that, in identifying the "what it is" and the "why it is" for propositions of the form "P is" as well as "S is P," Aristotle made the point that those predications of a subject that serve as the ultimate basis for scientifically explaining other attributes of that subject are those predications that make the subject the kind of thing it really is. This follows from the fact that definitions, which express the essence of the subject, are the first principles of demonstration. But Aristotle's application of this point to the case of propositions asserting the existence of attributes inherent in the subjects of the sciences (eclipse and harmony) made clear that in a sense "why is S a P?" is to be answered not only by a definitional predication of S but also by the answer to the question "what is P?"—that is, by the definition of P.[17]

After reminding us of this identification, Aristotle proceeds to offer an account (λόγος) of it. As Barnes points out, because the account that Aristotle proceeds to offer cannot be considered an argument for identifying the "what it is" and the "why it is," by positing a "harsh ellipse" we can take the λόγος Aristotle presents as showing how a definition can, in some sense, be demonstrated.[18] Aristotle begins this account by sketching various ways in which the essence of a thing can be related to the cause of the existence of that thing. Aristotle assumes that the existence of anything studied by a science has some cause identified by that science and remarks that this cause can be the same as that thing or different from it. I have already presented my interpretation of this distinction: for Aristotle the distinction between the things for which the cause of the existence is the same as those things and the things of which the cause of the existence is something different is the distinction between the things considered a basic substrate by a science (what I have called epistemic substances) and those attributes that can be demonstrated to inhere in some subject studied by a science. The cause of such an attribute is its subject, since an atomic premise on the basis

17. See chap. 2, sec. A.
18. Barnes, *Aristotle's "Posterior Analytics,"* 207.

of which the predication of that attribute of its subject is demonstrated will be derived from the definition of the subject. I take Aristotle to be appealing to this distinction in the present passage, when he distinguishes between entities of which the cause is the same and entities of which the cause is different.[19]

After distinguishing entities whose cause is different from themselves from those entities for which this is not so, Aristotle notes, "it is either demonstrable or indemonstrable." The grammatical subject of "is demonstrable or indemonstrable" (ἀποδεικτὸν ἢ ἀναπόδεικτον) is "the cause" (τὸ αἴτιον).[20] Aristotle is distinguishing two kinds of cause that are identified in explaining the existence of something. In some cases the cause by which one explains some fact is itself a brute fact, not such as can be rendered intelligible through demonstration. An example of this would be the slitting of an animal's throat considered as

---

19. A similar interpretation of the distinction in 2.8 and 2.9 between what has a cause different from itself and what does not is offered by M. Deslauriers: "if an item which has an αἴτιον is one of which some account (perhaps indemonstrable) can be given, then it is true to say that certain items 'explain themselves' in the sense that one can provide an account which will be both a description and an explanation. In other words, since the αἴτια of some things are displayed not in the middle terms of demonstrations (or only incidentally as middle terms), but in the predicates of identity statements, some things are identical with their αἴτια and some are not" ("Aristotle's Four Types of Definition," *Apeiron* 23 [1990]: 2 n. 2).

On the other hand, McKirahan (*Principles and Proofs*, 199–201) takes 2.8–10 to be concerned with neither subjects nor attributes but, rather, with *relations* between subject and attribute. Thus, thunder is a relation between a certain noise and a cloud, and so forth. He interprets the distinction here under consideration as between those relations that are caused by a more basic relation and those that are not. Thus, where I identify demonstrable attributes, complex subjects, and beings like eclipse and thunder, which are discussed in 2.8 and 2.10, McKirahan takes them to be three separate kinds of beings. This complicates the ontology of the *APo.* enormously, to little advantage. Further, Aristotle nowhere within the *APo.* considers relations as such; if McKirahan is correct, one would expect them to appear in *APo.* 1's account of the kinds studied by the sciences or at least in the aporetic chapters preceding 2.8–10. Third, Aristotle himself identifies the kind to be defined and the attribute demonstrated to hold of a subject, when he calls the eclipse a loss of light (2.8.93a23) and thunder a noise (2.8.93b9, 11). McKirahan (*Principles and Proofs*, 200, 203) takes Aristotle to be expressing himself sloppily, but the eclipse *is* a (certain) loss of light, and the thunder *is* a (certain) noise.

20. The only other grammatical possibility is to take the subject to be τί ἐστι. See J. Tricot, *L'Organon d'Aristote, traduction nouvelle et notes*, 2d ed. (Paris: J. Vrin, 1979), 188. To do so is in accordance with the general tenor of Philoponus' interpretation (*In Aristotelis*, 365, 14–36), according to which Aristotle is here introducing a distinction between a formal definition, which cannot be demonstrated, and a material definition, which *can* be demonstrated by taking the formal definition as a middle term. In secs. A and B I have discussed the problems involved in Philoponus' line of interpretation.

the cause of its death; there is a necessary connection between the slitting and the death that a demonstration could make intelligible (see 1.4.73b13–16), but it is not demonstrable that any animal or class of animals has its throat slit.[21] If the explanatory fact cannot be an object of scientific understanding, neither can the fact it serves to explain (cf. 1.2.71b21, 29–33). But, Aristotle points out, if the cause of the existence of some attribute is demonstrable, the existence of that attribute, too, can be demonstrated, through a universal first-figure syllogism whose middle term exhibits the cause of that attribute's being predicated of some subject.

Aristotle again faces the question: Could any definition be demonstrated in this way?

> One way to do this would be that which we just now examined, to demonstrate [what something is] through the other "what it is" statement. For when one is demonstrating what something is, it is necessary that the middle term be what something is, and when one is demonstrating [the existence of] properties, a property. So in the one case one will prove it, but one will not prove the other one of the identifications of essence by means of the same thing. (93a9–13)

The most obvious form that such a demonstration could take is that of a first-figure syllogism whose conclusion is a predication constituting the definition of some entity and whose middle term is some attribute that converts with both extremes. In *Posterior Analytics* 2.4 Aristotle has already considered this sort of "demonstration" and argued that it could not constitute a true explanatory demonstration of the fact that a certain predicate expresses the essence of some subject. Aristotle here refers back to that earlier discussion and remarks that, while such a syllogism at best has value in a dialectical (λογικός) context, it is not the sort of demonstration required.[22]

---

21. My interpretation is an alternative to that of Ross (*Aristotle's "Prior and Posterior Analytics,"* [Oxford: Clarendon Press, 1949], 629), Barnes (Aristotle's *"Posterior Analytics,"* 207–8), and Landor ("Aristotle," 130), for on their accounts *demonstrable* is here given a new and unprecedented sense according to which a demonstrable cause is one for which what is caused can be demonstrated to follow from the cause.

22. I here follow Ross (*Aristotle's "Prior and Posterior Analytics,"* 629–30) as taking 2.8.93a9–15 to be dealing with the kind of purported demonstration of a definition that had been discussed and rejected in 2.4. This is in contrast to the interpretations of Philoponus (*In Aristotelis,* 364–67), Tricot (*L'Organon,* 189), and Aquinas (*In Aristotelis,*

Aristotle then begins his discussion of what he takes to be a legitimate kind of demonstration of a definition. He approaches the problem in two stages. First, he clarifies the relationship between knowledge (γνῶσις) of what something is and knowledge that it exists.

> Just as we inquire about the why (τὸ διότι) after we have grasped that something is the case (ἔχοντες τὸ ὅτι), and (even though sometimes these become clear at the same time) it is not possible to know the reason why something is the case prior to knowing that it is the case, it is clear that it is likewise impossible to know the essence of a subject without knowing that it exists (ἄνευ τὸ ὅτι ἔστιν). For if one is unaware of whether something exists (ἀγνοοῦντας εἰ ἔστι) it is impossible to know what it is.
>
> Sometimes we have a grasp of whether something exists in an accidental manner, and sometimes we grasp something of the thing itself. (For example, something of thunder itself is that it is some noise in the clouds, of an eclipse, that it is a privation of light, of a human, that it is some animal, and of soul, that it moves itself.) But of all of those things of which we know only in an accidental manner that they exist (ὅτι ἔστιν), it is inevitable that we have no grasp at all of the "what it is" (τὸ τί ἐστι), for we do not know that those things exist (ὅτι ἔστιν); to inquire what something is without having a grasp of whether it exists (ὅτι ἔστι) is to inquire about nothing. But in the case of all of those things of which we have a grasp of something of the thing itself, it is easier to inquire what it is. So that the better a grasp we have of the fact that something exists, the better disposed we are towards what that thing is, too (ὥστε ὡς ἔχομεν ὅτι ἔστιν, οὕτως ἔχομεν καὶ πρὸς τὸ τί ἐστιν). (2.8.93a16–29)

Aristotle here returns to the issue of the relation between knowledge of existence and knowledge of essence, which he had discussed in 2.7. Aristotle had there indicated that one of the reasons for the incoherence

L.II,l.vii,470–72), according to which at 93a9–15 Aristotle is offering a kind of demonstration of a definition different from those hitherto discussed. There is too little in the passage to support the view that Aristotle is discussing the demonstration of a definition displaying one kind of cause by means of a definition expressing another kind of cause. Further, this line of interpretation, when developed, encounters those difficulties I indicated in secs. A and B.

of the notion of a demonstration of the essence of a kind is that such a demonstration would necessarily presuppose the existence of the kind, which existence in turn must be demonstrated on the basis of the essence of that kind. We have noted that Aristotle's reasoning was dialectical, since it ignored the teaching of *Posterior Analytics* 1 that the existence of some things is to be assumed, not demonstrated. Here Aristotle presents a related aporia, reminiscent of the *Meno* problem. Aristotle reiterates that we need to know that there is such a thing as an S before we determine what an S is. But he also indicates that one can determine that an S exists only if one knows what one is looking for, that is, what an S is (2.8.93a26–27; cf. *Meno* 80d5–e5, *APo.* 1.1.71a24–30). The dilemma is resolved by recognizing that one's cognitive grasp of the essence or the existence of a kind may be only partial. One may begin a successful inquiry into the essence of a kind on the basis of a grasp of the kind's existence that falls short of scientific understanding of that existence. Aristotle distinguishes four possible ways in which we may be related to the object of inquiry at the beginning of an investigation into its essence. In order of increasing adequacy they are (1) being ignorant of its existence, (2) knowing of its existence in an accidental manner, which involves no grasp of the essence, (3) knowing something of the thing itself, and (4) knowing both existence and essence. Aristotle indicates that "the better a grasp of the fact that something exists, the better disposed we are towards what that thing is, too" (2.8.93a28–29).

We will soon turn to the issue of how one is to understand each of these four ways of being disposed toward the object of investigation. But first we need to determine the meaning of Aristotle's correlation of awareness of existence and awareness of essence.

When Aristotle classifies and ranks the differences in how existence is grasped, his concern is not with various kinds of entities, some of which exist more than others (as is arguably the case when Plato distinguishes various cognitive states in *Republic* 5 and 6), but is, rather, with the various *ways* in which one and the same entity can be conceptualized. The fact that Aristotle refuses to consider "existence" a genus (*APo.* 2.7.92b14) indicates that he refuses to take it to be an independent attribute of a thing. Rather, the assertion that some kind S exists has cognitive content only by virtue of the cognitive content of the isolated subject S. This is because someone's assertion that an S exists is the assertion that certain attributes (those which the person takes to be necessary and sufficient conditions for being an S) are in fact predicated of some single subject. It is to say that there really is a thing in the world with those attributes that are taken to be the *meaning* of "S." To employ an example introduced by Aristotle in another context (*Metaph.* 7.4.1029b27–28), if I give the term *cloak* the meaning "what is pale and human," then for me to say that a cloak exists is nothing other than to say that a pale human exists.[23]

Consequently, the adequacy of one's awareness that S exists will vary in relation to the adequacy of one's awareness of what it is to be an S, which in turn will be more adequate the closer what one takes to be the meaning of "S" comes to the real essence of S. Limiting cases will be the complete ignorance of one for whom "S" is meaningless (or denotes something completely different from that which it denotes for all other speakers of the language) and the scientific understanding of one who fully grasps the real essence of S.[24] The intermediate cases will be those

23. Owen ("Aristotle on the Snares of Ontology," in *New Essays on Plato and Aristotle*, ed. R. Bambrough [New York: Humanities Press, 1965], 69–87; reprinted in *Logic, Science, and Dialectic*, ed. M.C. Nussbaum [Ithaca, N.Y.: Cornell University Press, 1986]) brought to light Aristotle's distinction between what Owen calls being* (being an X, where X is the essence of the subject) and being** (existence). He distinguished a third kind of being for beings such as time and place, that are posited within some theoretical account. For such things to be is for them to have the sense given to the terms referring to them within some account (90–92).

Upton ("The If-It-Is Question in Aristotle," *Ancient Philosophy* 11 [1991]: 326–29) has built on this account in an attempt to make clear what sort of being is ascertained for those things of which we have a partial grasp of existence. Upton applies Owen's analysis not only to theoretical posits but also to everyday objects that are studied by the sciences. "For example, when Aristotle asks the scientific if-it-is question of man, which should direct us to a grasp of man's causal what-it-is or causal *ousia*, he can rely on the prescientific instantiation awareness, or to put it more correctly, on the prescientific inductive acquaintance with particular instances, as the basis for the scientific question 'Does man exist as a certain kind of animal or not?'" (328) I agree. But I would add that this partial scientific account of existence need not be taken as having a logical form different from that of the assertion of existence asserted at what Upton calls "the ordinary level." An ordinary, or prescientific, assertion of existence would not be utterly "vague and unarticulated" (318). Rather, it would be to the effect that whatever the speaker takes to be necessary and sufficient conditions for the being in question are all exemplified. The same will be true for partial scientific accounts of existence. The difference is that, in the second case, what the speaker takes to be necessary and sufficient conditions for the being in question more adequately captures the being's essence.

24. This should make clear why *APo.* 2.8 has recently been mined for an Aristotelian semantic theory (see esp. Bolton, "Essentialism"). It is one of the few extended passages in which Aristotle is not primarily considering things themselves, but, instead, the different ways in which we can conceptualize the same things, or, what comes to the same thing, the different meanings that can be given to the terms that refer to things. Aristotle's

in which one knows something about the kind in question but in which the conceptual content of the kind in question is not an expression of the kind's real essence. One who is so related to the object of inquiry could express what one takes the object to be, but such an expression will not be a true definition, putting forward the real essence of the kind. Rather, it will be a nominal definition, expressing the *meaning* of the word referring to the kind in question.[25]

What meaning do we assign to each of the four kinds of grasp of existence: ignorance, accidental grasp of existence, grasping "something of the thing itself," and scientific knowledge of existence and essence? This is the subject of a long-standing controversy.[26] For Zabarella[27] one cannot grasp "something of the thing itself" without grasping the whole essence of the thing in question, and thus the third and fourth ways of being disposed toward the existence of a thing are to be identified. This interpretation has some support in Aristotle's claim that it is the grasp of "something of the thing itself" that makes possible a demonstration of the essence of that thing (2.8.93a27–28). As Ackrill points out, on this interpretation the present passage is not concerned with the various stages of the inquiry into a thing's essence; he is, rather, giving an account of what must be grasped in order that the inquiry be complete. I find this an unlikely interpretation of the text. Later in the chapter Aristotle does present a discussion of the progress from an apprehension of a thing through one of its nondefinitional

primary concern here is not the development of a semantic theory (though his suggestions are cogent and coherent and therefore might be profitably employed to form the basis of a semantic theory). Rather, an explication of the different ways in which one might be disposed toward the existence of something is necessary for the solution of the puzzle concerning the possibility of a demonstration of a καθ' αὑτὸ συμβεβηκός predication, which, in revealing the cause of the existence of the predicate, would make possible a scientific definition of the predicate. The puzzle can be solved only by distinguishing a preliminary grasp of what something is from a complete grasp of its essence. This is because such a demonstration consists in showing why there are things with those attributes that constitute one's initial grasp of the kind in question.

25. That an expression of an initial prescientific grasp of a thing, on which an inquiry into its nature will rest, will be a nominal definition is confirmed at 2.10.93b29–35. The distinction between the content of a nominal definition and a real essence corresponds to McKirahan's distinction between a "superficial essence" and a "deep essence."

26. A clear and concise treatment of the issues and major positions taken in this controversy can be found in J.R. Ackrill, "Aristotle's Theory of Definition: Some Questions on *Posterior Analytics* II 8–10," in *Aristotle on Science: Proceedings of the Eighth Symposium Aristotelicum*, ed. E. Berti (Padua: Editrice Antenore, 1981), 371–76.

27. J. Zabarella, *Opera Logica* (Hildesheim, Ger.: Olms, 1966), 1112–20.

attributes to an apprehension through a definitional attribute (93a35–b7), and this is presented not as a new topic but, rather, as having its basis on what has preceded.

The alternative is to interpret Aristotle's meaning as sketched earlier: the states of awareness of the existence of a thing that are intermediate between complete ignorance and scientific knowledge can act as stepping stones to attaining scientific knowledge of essence. The ease of the inquiry is determined by the extent to which the meaning of the term denoting the thing in question (as expressible in a nominal definition of that term) corresponds to the essence of the thing. The challenge of the interpreter of this passage is to determine precisely the nature of the two states of awareness of a thing's existence that are short of scientific knowledge.

Aristotle distinguishes accidental knowledge of a thing's existence from complete ignorance but indicates that it is of little if any use in scientific inquiry. What would an example of this be? Someone wholly ignorant of biology might be in the company of another who relates that she saw lemurs in her travels. The listener has some grasp of the fact that there are lemurs, but, if asked to give some account of what lemurs are, he could adduce only such accidental attributes as "a lemur is some entity that so and so saw in her travels." This attribute is accidental in the strongest sense: there is no necessary bond between the subject and the predicate.[28]

On the other hand, lemurs have some attributes that hold of necessity. The more of these attributes the listener learns of, the farther along the way to grasping the nature of lemurs the listener will be. To that extent, and to the extent to which the listener recognizes the causal relationships among these attributes (and hence approaches a grasp of the definition, which expresses the essence or ultimate cause of the attributes necessarily belonging to the kind), the sentence "Lemurs exist," which has been believed all along, has more determinate cognitive content. The sentence thus has a different sense for different people, though it has one privileged sense for all biologists.

Still to be considered is the third way in which we might be aware of a thing's existence: grasping "something of the thing itself." Unaccounted for in the other three kinds of awareness is the situation in

28. See the sense of "accidental" distinguished at *Metaph.* 5.30.1025a14–30. Ross presents an example of accidental knowledge similar to mine (*Aristotle's "Prior and Posterior Analytics,"* 630–31).

which one grasps one or a number of necessary attributes of a kind but still falls short of a grasp of the essence of the kind. If this account of an accidental grasp of the existence of a kind is correct, and if Aristotle's classification is comprehensive, it stands to reason that any such situation would be a case of "grasping something of the thing itself." I do take this to be the correct interpretation. But this is denied by Aquinas, Ross, and Barnes,[29] who take a grasp of "something of the thing itself" to necessarily be a grasp of a definitional attribute of a thing, excluding necessary but nondefinitional attributes, such as τὰ ἴδια, the nondefinitional attributes peculiar to the kind in question, and the καθ' αὑτὰ συμβεβηκότα. Aristotle's examples of a grasp of "something of the thing itself" at 93a22–24 are indeed all examples of a grasp of some definitional attribute. But, as Ackrill points out,[30] Aristotle would surely grant that a grasp of certain nondefinitional attributes of a kind, such as a property or demonstrable attribute, can greatly facilitate the inquiry into essence. Further, as we shall see, in 2.8.93a39–b7 Aristotle presents a sketch of a successful inquiry into the essence of a lunar eclipse that begins with a grasp of a nondefinitional attribute of the eclipse: "no shadows are cast by the Moon even though it is full." Clearly, one starting with such an understanding of the lunar eclipse has more work to do than someone who recognizes that it is a blocking of light. But at the same time such an investigator knows more about the eclipse than one knowing an entirely accidental attribute such as "the eclipse is what the students wish to see through their telescope."[31]

To summarize, Aristotle's main point in this passage is that we can learn what a thing is and that this is easier to the extent that a pretheoretical concept of a thing approaches its real essence. The examples Aristotle employs make clear that *these* remarks hold in regard to both

29. Aquinas, *In Aristotelis*, L.I,l.ii,475; Ross, *Aristotle's "Prior and Posterior Analytics,"* 630; Barnes, *Aristotle's "Posterior Analytics,"* 209.

30. Ackrill, "Aristotle's Theory," 373–74.

31. That the inquiry into what something is, is aided by a grasp of an attribute of that thing (essential or not) is also stated at *EE* 2.1.1220a15–22: "We must inquire in the manner of everyone engaged in other inquiries, who have some grasp of what they seek (ἔχοντές τι), so that from what is said truly but not clearly, we must try to have a true and clear grasp of things. But now we are in the state that we would be in if we were to say about health that it is the best disposition of the body, or if we were to say that Corsicus is the palest of those in the agora. For in these cases we do not know what each of these is, yet to be in this state is a step towards (πρὸ ἔργου) knowing what each of these is" (reading εἰ λέγομεν at 1220a18, with Russell). While the remarks at *APo.* 2.8.93a21–29 concern scientific inquiry, the remarks in the *EE* concern all inquiry.

the things that have no different cause, theoretical simples such as human and soul, and what has a different cause as well, attributes such as the eclipse. Scientific knowledge of the existence of the former will be determined through inductive observation and through dialectical discussion (for which an account is given in the *Topics*). But scientific knowledge of the existence of the latter will come about through the sort of inquiry and demonstration Aristotle is about to describe. For, in the passage that follows, Aristotle considers inquiries into the essence of demonstrable attributes of the subjects of the sciences.[32]

D. The Structure of the Demonstrations of the Existence of "That of Which There Is a Different Cause"

Aristotle begins by considering how demonstration can be employed to determine the true essence of a lunar eclipse.

> Let this be the first example of the sort of inquiry for which we grasp something of the "what it is." Let eclipse be A; Moon, C; blocking of light by Earth, B. Then to inquire whether it is eclipsed or not is to inquire whether there is B or not. And this is not different from inquiring whether there is an account of it (εἰ ἔστι λόγος αὐτοῦ). And if this should exist, we say that A exists too. Or, to take another example, of which side of the contradiction is there the account (ἢ ποτέρας τῆς ἀντιφάσεώς ἐστιν ὁ λόγος)? Is it of having or of not having two right angles? When we learn this, we know what is the case and why it is the case (τὸ ὅτι καὶ τὸ διότι) simultaneously, should this come about through unmediated premises. If we do not learn the fact in this way, we know what is the case, but not why it is the case. (2.8.93a29–37)[33]

At 2.2.89b7–8 Aristotle had identified the inquiry into whether a subject exists and the inquiry into whether there is a cause (or middle term). One of his examples of an inquiry into the cause of a thing was

32. Insofar as Aristotle here envisages the inquirer to attain real explanatory demonstrations of the inherence of these attributes in their subjects, he is apparently also presuming that the one making the inquiry has already secured the first principles of the sciences dealing with those subjects.

33. My understanding of this difficult passage, as outlined here, owes much to the help of A.P.D. Mourelatos and Sarah Broadie.

the inquiry into the middle term of the existence of the Moon (90a5–7). Aristotle thus indicated that the being of even a substance, as well as that of a demonstrable attribute, can be said to have a middle term as a cause. This indicates that τὸ μέσον (the middle) can signify even the sort of cause that will not be expressed as a middle term of a demonstrative syllogism. In the case of the being of a substance, the middle term will be the essence of that substance. So, when in the present passage Aristotle identifies the inquiry into whether something exists with the inquiry into whether there is a λόγος of it, he is not identifying the inquiry into whether a subject exists with the inquiry into whether the existence of that subject is demonstrable. Rather, the inquiry into whether a subject exists is the inquiry into whether, from the point of view of the sciences, the subject has *some* explanation or cause, even if that explanation is the subject itself.[34] The scientific inquiry into whether an eclipse exists will be the inquiry into whether there is an explanation for the existence of eclipses. But, insofar as the real essence of the eclipse is as yet unknown, this inquiry will take the form of an inquiry into whether those attributes that are expressed in one's *nominal definition* of the eclipse are caused. Thus, given that one understands a lunar eclipse as a blocking of light by Earth, the inquiry into whether a lunar eclipse exists or not is equivalent to the inquiry into whether Earth blocks light from the Moon or not. In the context of astronomy this is in turn to be identified with whether there is some cause of Earth's blockage of light from the Moon.

Aristotle's comments so far hold for the inquiry into the existence of anything. But Aristotle proceeds to focus on a special case: that in which both the existence and the cause of the existence are secured simultaneously "through unmediated premises." This is the case in which one acquires scientific understanding of the existence of an attribute, what "has a different cause," through an explanatory demonstration of the fact that a certain attribute is predicated of a certain subject. That the rest of the chapter is concerned with a method of under-

34. On the sense of λόγος here, see Ross, *Aristotle's "Prior and Posterior Analytics,"* 628; and Mansion, *Le jugement*, 185. Note that, in identifying the inquiry into whether something exists with the inquiry into whether it is caused, Aristotle is making clear that he is here concerned with *scientific* inquiry into existence. For every science there are some things whose existence is not to be explained on the basis of the first principles of that science. Though such a thing does exist in reality, from the perspective of that science its existence is a brute fact insofar as it is indemonstrable. Hence *that science* will be unable to affirm the existence of the thing or kind in question.

standing the existence of "what has a different cause" through an explanatory demonstration is confirmed by the fact that all of the examples Aristotle employs from here on are of "what has a different cause."

At this point in our exegesis it is useful to keep in mind the distinction made in 1.1.78a22–b4 between a truly explanatory demonstration (which Aristotle calls "of the reason why" [τοῦ διότι]) and a demonstration of the fact (τοῦ ὅτι). The former kind of demonstration is the mainstay of the sciences: such demonstrations are grounded entirely on atomic predications that make clear *why* the conclusion holds. A demonstration of the latter kind, like the former kind, has as its premises objects of scientific knowledge. But because these are not necessarily atomic predications, it is possible that the conclusion of such a demonstration states the cause of one of the premises. Such a demonstration serves as a proof that the conclusion is the case but does not explain why it is the case. Every explanatory demonstration can also function as a proof that the conclusion of the demonstration is true, but not every proof that a conclusion is true serves to explain that conclusion.[35]

Aristotle makes precisely this point at 93a29–37. He indicates that, if grounded on unmediated premises, a demonstration that there exist together those characteristics that we take to constitute a certain kind will not only serve as a vehicle for determining that that kind does exist but will also at the same time reveal why it exists: it will be both a demonstration τοῦ ὅτι and a demonstration τοῦ δίοτι. But, if the demonstration of the existence of some attribute rests on premises that have not yet themselves been demonstrated, the demonstration has not revealed all the root causes of the fact; for this reason it cannot yet qualify as a true demonstration τοῦ δίοτι. In such a case we may be able to say that we know that some attribute exists or that some fact is the case (for demonstration rests on premises that are accepted as sanctioned by science), but we cannot be said to be able to explain *why* that fact is true.

Aristotle gives two examples. The geometric proof that every triangle has the sum of its angles equal to that of two right angles, if grounded on the first principles of geometry, will not only prove that this is the case for every triangle but will show why it is the case. Similarly, as I have noted, if we understand the lunar eclipse to be a block-

35. This is pointed out by G. Patzig, "Erkenntnisgründe, Realgründe, und Erklarungen (zu *An. Post.* A 13)," in *Aristotle on Science: Proceedings of the Eighth Symposium Aristotelicum*, ed. E. Berti (Padua: Editrice Antenore, 1981), 145.

ing of light by Earth (which understanding of the eclipse will be expressed in a nominal definition of *eclipse*), then once we have an explanatory demonstration that this blocking of light by Earth does indeed sometimes occur in respect to its subject, the Moon, by virtue of substituting coextensive terms, we can be said to have demonstrated that the Moon does indeed sometimes suffer a lunar eclipse. This substitution will follow the demonstration proper, which will not itself include the term *eclipse*.[36]

To anticipate a bit, we can already see how one could demonstrate that a καθ' αὑτὸ συμβεβηκός attribute inheres in a subject, even though that attribute is itself neither a subject of the sciences nor either a genus or differentia of such a subject. One would demonstrate that the definiens of a nominal definition of an attribute inheres in the subject of that attribute. This would be possible if all of the terms employed in that demonstration are either subjects of the sciences or terms occurring in the definitions of such subjects.

But, as I indicated, there is no necessity that the "something of the thing itself" that is grasped as the nominal definition of a kind be an element of the essence of that kind. Aristotle next presents an example of this in the case of the lunar eclipse and shows how even one in this situation can successfully resolve the inquiry into why the kind exists.

> Let Moon be C; eclipse, A; the inability to create a shadow during the full moon even though nothing evident is between us, B. Then if B (the inability to create a shadow even though nothing evident is between us) is predicated of C, and A (to have been eclipsed) is predicated of this, it is clear *that* the Moon is eclipsed, but it is not yet clear *why* it is eclipsed, and we know *that* an eclipse exists, but we do not know *what* it is. But once it is clear that A is predicated of C, to inquire why it is predicated is to inquire what B is, whether a blocking, turning, or quenching of the Moon. This (what B is) is the account of the other extreme, e.g., in the case of these terms, of A, for an eclipse is a blocking by Earth. (2.8.93a37–b7)

36. This is because none of the demonstration's unmediated premises could include a term like *eclipse*, defined as a certain conjunction of demonstrable properties. Ross (*Aristotle's "Prior and Posterior Analytics,"* 77–78) recognizes that the true explanatory demonstration described in 2.8 will not itself include the term whose definition that demonstration serves to explain but apparently does not see that Aristotle thereby presents a means of demonstrating any καθ' αὑτὸ συμβεβηκός predication.

Aristotle is here imagining a case in which the one investigating the phenomenon of the lunar eclipse initially grasps that phenomenon only as the inability of the Moon to cause the shadow of a body between it and the ground to be cast, even though the Moon is full and there is no other body between the Moon and that which is to cast the shadow.[37] In response to the question "Why are there eclipses of the Moon?" the following deduction S1 can be given:

(i) The Moon (C) is such as to sometimes not cause shadows to be cast even if there is nothing between it and the body, and even though it is full,
(ii) for something to not cause the shadow of a body to be cast even if there is nothing between it and the body, even though it is full, is for it to be eclipsed,
so, (iii) the full Moon is sometimes eclipsed.[38]

This is a demonstration τοῦ ὅτι but is not a demonstration τοῦ διότι. Although this deduction rests on premises that are true and accepted as sanctioned by the sciences (and hence is such as to grant knowledge *that* the conclusion is true), its premises are not such as not to demand explanation, for which reason it does not reveal *why* the conclusion is true. It is the case that premise ii does not require demonstration and can be accepted without further agreement, for this premise is a nominal definition, initially fixing the signification of the term *eclipse* for the one conducting the inquiry. Premise i, on the other hand, although verifiable empirically, is not unmediated and hence *is* such as to demand further explanation. One therefore requires an explanatory demonstration, which will display the cause of this phenomenon. The middle terms of the demonstration are not to be determined a priori, for it is

37. This last condition must be added although it is unstated by Aristotle, for, otherwise, the New Moon would be considered to be eclipsed.

Aristotle's example of a nominal definition of the lunar eclipse is a bit artificial; it is unlikely that anyone would understand an eclipse as characterized by this phenomenon alone. Bolton ("Definition," 134–36) makes this point but nonetheless has a clever suggestion regarding the circumstances under which one might take an eclipse to be the phenomenon of the Full Moon's not causing shadows to be cast. He imagines a hunter in the woods who knows that it is the night of the Full Moon and that the weather is clear yet notes the absence of the normal patterns of light and shadow that facilitate his hunting.

38. For the sake of simplicity, here and in the syllogisms that follow, I do not explore the issue of how temporal qualifications such as "sometimes" are to be handled within the framework of Aristotelian logic.

equally conceivable that the phenomenon that is recognized as an eclipse is caused by a blocking of light, a turning of the Moon's bright side, or a quenching of some fire. Its determination must come about through empirical research. Through such research we will discover that at the time of the lunar eclipse Earth is between the Sun and the Moon. We then infer that, at the time in which the Full Moon is unable to cast shadows, though nothing is between it and Earth, Earth is blocking the light of the Sun from the surface of the Moon. Once this is recognized the lunar eclipse can be fully explained[39] through the following pair of explanatory demonstrations.

S2: (i) The Moon is such as to sometimes have Earth come between it and the Sun,
(ii) what is such as to have Earth come between it and the Sun will sometimes be "full" and have the sunlight blocked from its surface,
so, (iii) the Moon is such as to sometimes be full and have the sunlight blocked from its surface.

S3: (i) The Moon is such as to sometimes be full and have the sunlight blocked from its surface,
(ii) whatever is full and has the sunlight blocked from its surface will not cause a shadow to be cast,[40]
so, (iii) the Moon is such as to sometimes be full and not cause a shadow to be cast.

All three premises of the demonstration are presumably demonstrable on the basis of atomic predications that are definitional of light and the celestial bodies.[41] Accordingly, these demonstrations would meet Aristotle's criteria for presenting an ultimate explanation of why it is the case that the phenomenon that the inquirer refers to as "lunar eclipse" does in fact occur.

---

39. At 93b6 the middle term is said to be the λόγος of the major term in the sense of "explanation," not "definition," the suggestion of Landor ("Aristotle," 130).

40. I am simplifying here; this premise will not be true for those bodies that are luminous by virtue of nonreflected light. But it could be demonstrated on the basis of astronomical premises that the Moon is not such a body.

41. Definitions of the astronomical bodies will often be definitions that hold of a single thing alone, as is the case for the Moon. This led Barnes (*Aristotle's "Posterior Analytics,"* 211) and Landor ("Aristotle," 131) to the view that the syllogisms sketched in 2.8 are

Aristotle presents a parallel sketch of the resolution of the inquiry into the nature of thunder.

> What is thunder? A quenching of fire in the cloud. Why does it thunder? Because the fire in the cloud is quenched. Let cloud be C; thunder, A; quenching of fire, B. Now B is predicated of C, "cloud" (for the fire is quenched in it), and A (noise) is predicated of this. And it is true that B is the account of A, the first extreme. Should there be another middle term besides this, it will be based on the remaining accounts. (2.8.93b7–14)

The phenomenon to be explained (thunder) is denoted by A, the very term that denotes noise, and it is this noise that is demonstrated to be in the clouds. Aristotle is imagining a case in which "noise in the clouds" is the "something of thunder itself" that is taken as the inquirer's initial grasp of what thunder is.[42] It therefore occupies the same place in the inquiry into thunder as "the inability to cause a shadow to be cast . . ." has in the inquiry into the eclipse. Insofar as we know empirically that there is noise in the clouds, the following syllogism S1′ is possible:

> (i) The clouds are such as to often be clouds with noise occurring in them,[43]

particular, having the form of Darii. But, if this is so, what the demonstrations make clear is not the existence and essence of the natural kinds thunder and lunar eclipse but, rather, the existence and essence of particular instances of these kinds. This would fall short of giving one scientific understanding of these kinds. It is important to recognize that definitions of kinds that have only a single member remain, qua definitions, definitions of *kinds* or species, for otherwise there could be no scientific knowledge of those kinds (*APo.* 1.18; *Metaph.* 7.15.1039b27–30, 1040a27–b2). (On this, see A.C. Lloyd, "Necessity and Essence in the *Posterior Analytics*," in *Aristotle on Science: Proceedings of the Eighth Symposium Aristotelicum*, ed. E. Berti [Padua: Editrice Antenore, 1981], 170–71.) Hence, demonstrations grounded on such definitions can still be considered universal and will retain their place in the exposition of a science.

42. This is a simplification, although, for my present purposes, an innocuous one. Aristotle understands "something of thunder itself" to be "a *certain* noise in the clouds" (ψόφος τις νεφῶν) (*APo.* 2.8.93a22–23). Bolton ("Definition," 137 n. 31) is no doubt correct in stressing that Aristotle believes that we take the familiar rumbling in the clouds to be thunder and that other noises in the clouds (such as the noise of winds) would not be classified as thunder, even by one with only a pretheoretical grasp of what thunder is. Cf. R. Sorabji, "Definitions," 216–17; Landor, "Aristotle," 130.

43. Presenting the premise in this way, and not as asserting that something is true of a particular cloud, preserves the universality of the demonstration. On this, see n. 42.

(ii) whatever is a cloud with noise occurring in it thunders,
so, (iii) the clouds are such as to often thunder.

Like S1, this syllogism fails to explain why there is such a thing as thunder, since the minor premise (i) itself demands explanation. This is remedied by the following demonstration, S2′:[44]

(i) The clouds are such as to often have fire quenched in them,
(ii) whatever has fire quenched in it makes noise,
so, (iii) the clouds are such as to often make noise.[45]

Both premises of this demonstration are presumably themselves demonstrable on the basis of the principles of physics and acoustics.[46] Here, as before, the existence of the phenomenon in question becomes an object of scientific understanding by virtue of a demonstration in which the term denoting the phenomenon in question is nowhere employed.[47]

Aristotle does not, of course, sketch in full the preceding demonstrations. Indeed, he does not sketch demonstrations at all; rather, in each case he presents one middle term, (*blocking*, explicated later as "block-

---

44. I here follow neither of the alternative syllogisms sketched by Ackrill ("Aristotle's Theory," 360) although my suggestion comes close to the second ("A noise is quenching of fire, quenching of fire is [occurs] in clouds, so, a noise is [occurs] in clouds"). I believe that my suggestion is an improvement in two respects. First, the demonstration as I see it is universal, as Aristotle believes it should be (*APo.* 1.24). Second, its temporal qualification takes account of the requirement that the premises of a demonstration be true at all times (cf. 1.8.75b33–36).

45. At 2.10.94a7–9 "the conclusion of the demonstration of the 'what it is' " of thunder is identified as "noise in clouds." Aristotle may well have in mind the conclusion of syllogism S2′.

46. I take these to be the "remaining accounts" (παράλοιποι λόγοι) referred to at 93b13–14. This is the interpretation of Aquinas (*In Aristotelis*, L.II,l.vii,477) as well: "Sed si oporteat accipere aliquod medium ad hoc demonstrandum, hoc assumetur *ex reliquis rationibus*, idest ex definitione minoris extremitatis, et aliarum causarum extrinsecarum. Cum enim subiectum sit causa passionis, necesse est quod definitio passionis demonstretur per definitionem subiecti. Et hoc patet in exemplo propositio. Quia enim luna est corpus natum sic moveri, idea necesse est quod obiiciatur certo tempore terra inter solem et ipsam." This is also the view of Ross (*Aristotle's "Prior and Posterior Analytics,"* 77–78). Cf. Mansion, *Le jugement*, 191.

47. Such a demonstration will be possible only given that the account of thunder is indeed the correct one. Aristotle's own account, given in *Mete.* 2.9, can also be expressed through a demonstration of the kind discussed in this chapter. Cf. the proposals in C. Freeland, "Scientific Method in Aristotle's Meteorology" (photocopy, 1986).

ing by Earth," and "fire being quenched") as making possible an account of the first, provisional middle term (*not causing a shadow to be cast* . . . and *noise*). But Aristotle is clearly giving an abbreviated account of a true explanatory demonstration of the phenomenon in question, which must be grounded on unmediated premises (cf. 2.8.93a6), and the syllogisms sketched here present the main structure that this syllogism would have to take.

Aristotle presents these demonstrations as illustrations of how there can be a demonstration of the essence of a kind that is not subject to the aporiai raised in 2.3–7. But what has Aristotle in fact presented? On the face of it Aristotle has presented demonstrations that there exists a certain conjunction of attributes, identified through a nominal definition as the referent of a certain term. He does not here explicitly describe how such a demonstration can be taken to have a definition as its object. It has as its conclusion an assertion of existence, not an assertion of essence. That this is so should not be surprising, since we have seen that in many of Aristotle's aporetic arguments against the possibility of there being a demonstration of a definition it was stressed that there is a sharp distinction between the kind of proposition that can be the conclusion of a demonstration (an assertion of existence) and the kind that is an ultimate premise of a demonstration (a statement of essence). What is surprising is that such a series of syllogisms is nonetheless presented as "of a definition."

Aristotle's intention is clarified by the way in which he words his conclusion at the end of 2.8 and in 2.9:

> That the "what it is" is assumed and becomes known has been said; so that there comes to be neither a deduction nor a demonstration of the "what it is." Nonetheless, it becomes clear through a syllogism and through a demonstration (δῆλον μέντοι διὰ συλλογισμοῦ καὶ δι' ἀποδείξεως). So that, in the case in which there is another cause, it is not possible to know the "what it is" without a demonstration, and there is no demonstration *of* it (ὥστ' οὔτ' ἄνευ ἀποδείξεως ἔστι γνῶναι τὸ τί ἐστι, οὗ ἔστιν αἴτιον ἄλλο, οὔτ' ἔστιν ἀπόδειξις αὐτοῦ), as we said when presenting the aporiai. (2.8.93b15–20)
>
> In the case of some things there is a different cause; in the case of others, there is not. So it is clear that among statements of what something is some are unmediated and are principles, in which

> case the existence and the "what it is" of the thing must be hypothesized or made clear in some other way (ὑποθέσθαι δεῖ ἢ ἄλλον τρόπον φανερὰ ποιῆσαι), which is the very thing the arithmetician does, for he hypothesizes what a monad is and that there is such a thing; but of those things that have a middle, and of which the essence has some different cause, the "what it is" is, as we have said, made clear through demonstration, but not by demonstrating the "what it is" itself (ἔστι δι' ἀποδείξεως . . . δηλῶσαι, μὴ τὸ τί ἐστιν ἀποδεικνύντας). (2.9.93b21–28)

Aristotle here makes a distinction between two different ways in which a demonstration may be related to a proposition p.[48] According to the first, the demonstration is said to be "of p" (where *p* is in the genitive case) and the one who is responsible for the demonstration is said to be demonstrating p (where *p* is in the accusative case). According to the second, the demonstration "makes p clear." Aristotle here employs the first idiom for cases in which p is the conclusion of the demonstrative syllogism.[49] The second idiom, "to make clear" (δηλῶσαι), is synonymous with "to make manifest" (φανερός ποιῆσαι); it is to allow something to be seen or known. To make p clear is a nontechnical notion that is more encompassing than either "to syllogize p" or "to demonstrate p," each of which denotes a particular way by which one can make p clear.

The aporiai of *Posterior Analytics* 2.3–7 were directed against the possibility of a demonstration's being "of a definition" in the sense of hav-

48. See Ferejohn, *Origins*, 148 n. 36.

49. The verb ἀποδεῖξαι (to demonstrate) and its cognate ἀπόδειξις (demonstration) follow the same syntactical rules as συλλογίζεσθαι (to syllogize, to deduce) and συλλογισμός (syllogism, deduction). The accusative object of both verbs signifies the conclusion of the deduction signified by the verbs (e.g., *APr.* 2.5.58a35: τὴν ἑτέραν πρότασιν τὴν λοιπὴν συλλογίζεσθαι; *APr.* 2.16.64b29: τὸ μὴ ἀποδεικνύναι τὸ προκείμενον) or the deduction itself (e.g., *APr.* 1.30.46a26–27). When ἀπόδειξις or συλλογισμός is modified by a genitive word or phrase, the genitives signify what the deductions reveal. This is usually the conclusion of the deduction. Thus, the συλλογισμός τοῦ ὅτι discussed at *APo.* 1.13.78a36–37 has as its conclusion the fact signified by the phrase τὸ ὅτι. It is in this sense that Aristotle in *APo.* 2.8–9 denies that there is an ἀπόδειξις τοῦ τί ἐστι. On the other hand, the genitive word or phrase modifying ἀπόδειξις or συλλογισμός sometimes signifies something that the deduction reveals without having it as its conclusion. See, for example, the deduction discussed at *APo.* 1.13.78a39–78b4, which is called both an ἀπόδειξις τοῦ διότι and a συλλογισμός τοῦ διότι but does not have the cause expressed in the conclusion. It is in this looser sense that at *APo.* 2.10.94a12–14 Aristotle refers to the sort of demonstration discussed in *APo.* 2.8 as both an ἀπόδειξις τοῦ διότι and a συλλογισμός τοῦ διότι.

ing a definition as a conclusion. In the present passage Aristotle reaffirms that this is impossible.[50] But Aristotle claims that a demonstration of the sort he has just sketched is nonetheless an indispensable step in the inquiry into the essence of certain kinds, those "of which there is a different cause." As such, it serves to make an essence clear, even though it is not of a definition, in the sense of having the definition as its conclusion.[51] But in the case of what does not have "a different cause"—that is, the epistemic substances[52]—no such demonstration is possible.

How does the kind of demonstration that Aristotle has sketched help to reveal the definition of the major term? One must first recall what a scientifically adequate definition is for Aristotle: it not only allows a speaker to correctly identify instances of a certain kind (which is the role played by a nominal definition), but it also reveals the *cause* of the existence of the kind in question. In the case of things that come to be and perish (like an eclipse or thunder), the definition will specify some event that is the efficient cause of the kind in question. In the case

50. On this see Landor, "Aristotle," 128. The reference to the these aporiai (ὥσπερ καὶ ἐν τοῖς διαπορήμασιν εἴπομεν [at 2.8.93b19–20]) shows that Aristotle takes at least some of these aporiai to be sound, having more than dialectical worth.

51. This is seen by Ross (*Aristotle's "Prior and Posterior Analytics,"* 78, 631–32) and Landor ("Aristotle," 124). It is not seen by Bolton ("Definition," 141), who objects to an interpretation of the "demonstration of an essence" along these lines: "The primary objects of geometry are points and lines (1.10.76b3ff). There is, however, simply no way to derive by demonstration the definition of triangle, say, simply from the definitions of point and line." Because Aristotle does not indicate that the demonstration making a definition clear has the definition in question as a conclusion, he need not claim that the definitions of what has a cause other than itself are in any way *derived* from the definitions of other things. Rather, a grasp of the meaning of the term is presupposed; the demonstration *explains* how it is that the term refers to real attributes. The definition of triangle is not derived or deduced on the basis of other definitions. Rather, starting with a definition giving the *meaning* of "triangle" (a three-sided plane figure), Aristotle envisages a demonstration showing that lines are the sort of thing that can form a three-sided plane figure, in this way explaining the fact that the "nominal definition" refers.

52. In commenting on *APo.* 2.9, Aquinas (*In Aristotelis*, L.II,l.viii,482) presents an interpretation of "the things of which there is no different cause," as things very similar to my "epistemic substances": "Tertio modo potest intelligi in quantum aliqua sunt, quae non habent causam in genere subiecto alicuius scientiae: sicut in genere numeri, de quo est arithmetica, est devenire ad unitatem, cuius in hoc genere non est accipere aliud principium. Et hic sensus concordat exemplo quod Philosophus subiungit, dicens quod arithmeticus supponit *quid est* unitas et *quia est.*" But Aquinas does not explain why the kind of demonstration that Aristotle has sketched in 2.8 is inapplicable in respect to this kind.

of eternal kinds (such as mathematical attributes), the definition will present only the formal characteristics of the kinds in question.[53]

With this in mind, it can be seen how the sort of demonstration Aristotle sketches in *Posterior Analytics* 2.8 makes it possible to resolve the inquiry into essence. Given an explanatory demonstration of the fact that this conjunction of predicates inheres in some subject, one who conceptually identifies some kind K with a certain conjunction of predicates will be able to say, "Now I know why there are K's." This is because, like all explanatory demonstrations, its middle terms would identify the causes of the existence of the major term. One would then be able to attain a scientifically adequate definition of K by conjoining the definiens of the pretheoretical nominal definition of K, which is demonstrated to exist, with the attribute of being caused by the referent of a middle term of that demonstration.[54] Thus, the definition of the lunar eclipse made possible by the demonstration of its existence is "a dimming of light on account of the blocking of Earth" (cf. 2.2.90a15–16). The definition of thunder made possible by the demonstration of the existence of thunder is "noise in the clouds on account of the quenching of fire" (cf. 2.10.94a5).[55]

But we have seen that, according to the theory of explanation presented in *Posterior Analytics* 1, not every kind studied by the sciences will be such as can have its existence demonstrated. The existence of the epistemic substances, or basic subjects studied by the sciences, will be assumed by the sciences studying them. Hence, the definitions of the epistemic substances cannot be revealed by this kind of demonstration. They, too, must be posited by the sciences as first principles. If, as I have argued, "what has no different cause" is Aristotle's term here for the epistemic substances, Aristotle is making precisely this point at the end of 2.8 and within 2.9. In no sense at all is there a demonstration of the definition of an epistemic substance, neither as the conclusion of a demonstration nor as merely "made clear" by a demonstration. It must

53. Cf. *Metaph.* 7.17.1041a28–32, and my discussion of this passage in chap. 2, sec. C.

54. Cf. Ross, *Aristotle's "Prior and Posterior Analytics,"* 77–78, 631–32; and Landor,"Aristotle," 129.

55. As we shall see, this interpretation of how a demonstration can reveal a definition is confirmed at 2.10.93b38–94a7, in which Aristotle claims that the one λόγος of thunder can be expressed as either a definition or a demonstration. At 2.10.94a12–13 the definition and the demonstration to which it is correlated are said to differ "in how the terms fall" (πτώσει). In both cases the nominal definition and the key middle term are presented, but there is a difference in the forms of the accounts in which they are presented.

be "hypothesized or made clear in some other way," just as the arithmetician assumes the existence of the epistemic substance of arithmetic, that is, the monad (2.9.93b23–25). Only those entities whose existence is logically entailed by the existence and essence of an epistemic substance (those with "a middle," or a cause, different from themselves [2.8.93b19, 2.9.93b25-26]) are such as can have their essences revealed through a demonstration of the kind Aristotle has sketched.[56]

E. The Kinds of Definitions: *Posterior Analytics* 2.10

This interpretation is confirmed in *Posterior Analytics* 2.10, which is something of an appendix to the inquiry into the demonstrability of definitions. Within it Aristotle reviews the different sorts of accounts that can be called definitions. He begins as follows:

> Since a definition is said to be an account of what something is (ὁρισμὸς δ' ἐπειδὴ λέγεται εἶναι λόγος τοῦ τί ἐστι), it is clear that one kind of definition will be an account of what the word means, i.e., another expression that has the force of a single word,[57] for example, what "triangle" means. In this case when we have a grasp of the fact that something exists, we inquire about why it exists. It is difficult to grasp in this way the things that we do not know to exist. The cause of this difficulty was stated earlier: that we do not know whether such things exist or not, except in an accidental sense. (2.10.93b29–35)

We recall that in 2.8.93a21–29 Aristotle argues that the ease with which one discovered the essence of something is determined by the extent to which we have a cognitive grasp of its existence. We have seen that he holds that the adequacy of one's cognitive grasp of a thing's existence is determined by the extent to which there is a correspondence between the meaning assigned to the term referring to the thing and the true essence of the thing. Because one's conception of the meaning of a term would be expressible through a nominal definition, the demonstration of the existence of the kind in question takes the form of

56. On this, see Ross, *Aristotle's "Prior and Posterior Analytics,"* 78, 633.

57. I am indebted to A.P.D. Mourelatos for this translation of λόγος ἕτερος ὀνοματώδης.

a demonstration that there inheres in some subject all of those attributes expressed in the nominal definition of that kind.

Here Aristotle makes explicit the fact that one's conception of what a thing is, those attributes that must be shown to exist in order for it to be shown that the thing exists, is expressed in a kind of definition: a nominal definition. (Here this is called "another expression that has the force of a single word.") Aristotle also confirms that the ease of our inquiry into why the thing exists is determined by the adequacy of that nominal definition.[58]

A nominal definition can be called a definition because it is, in a sense, an account of what something is, where the *is* is the copulative ἐστι.[59] If a nominal definition of a kind is not to be utterly misleading,[60] that kind will indeed have predicated of it those attributes that are identified within that nominal definition. But a nominal definition need not express the real ontological constitution of a kind, its essence.[61]

Aristotle appends to the previous passage a parenthetical remark: "An account (λόγος) is one in two ways: one way is by being tied together, like the *Iliad*, and one by making clear one thing in respect to

---

58. I take the ὅπερ (in this case) at 93b32 and the οὕτως (in this way) at 93b33 to indicate that the way in which one has a partial grasp of a thing's existence is by having a nominal definition, as described in 93b29–32. It is also possible to give the οὕτως here the sense of "by virtue of a knowledge of its real cause" (διὰ τί ἔστιν). On the relative merits of the two ways of interpreting οὕτως, see Ackrill, "Aristotle's Theory," 375–76.

59. Because a nominal definition of a kind is required for the investigation into the essence of the kind to proceed, it has something of the status of a first principle. Thus, at *APo.* 1.1.71a15 the τοδὶ σημαίνει of a derivative entity (triangle) is identified as the kind of thing that must be assumed prior to geometrical learning and teaching. But Aristotle's terminology is imprecise in this passage, for in the next line (71a15–16), when Aristotle is indicating that, for an epistemic substance like a monad, both a definition and an assumption of existence are necessary, he also calls the requisite definition an expression of τί σημαίνει, though in this case the definition would be an expression of the real essence of the definiendum. On this passage, see chap. 3, sec. B.

60. If one's investigation of a kind K begins with a nominal definition of K in which attributes are predicated of K that are never found together (e.g., "a horse is a hoofed quadruped with a single horn growing from its forehead"), it is clear that one's investigation of K will be as difficult, if not more so, than the investigation of one with merely "accidental" knowledge of K. If there are some entities that have predicated of them those attributes conjoined in one's nominal definition of K, but K (the kind that others in the same linguistic community refer to as "K") is not among them (for example, "a horse is a carnivorous quadruped"), one will be likewise ill equipped to determine the real essence of the kind most people refer to as a "K."

61. The sort of being that a scientifically adequate definition must express is what in *Metaph.* 5.7.1017a7–8 and 1017a22 is called καθ' αὑτό being, distinguished from "accidental" being.

another in a non-accidental way" (2.10.93b35–37). Insofar as the bond between subject and predicate is not accidental, the thing and its predicate are a true unity, that is, the predicate expresses what the thing really *is*.[62]

The next kind of definition distinguished here is a scientifically adequate definition of a kind "of which there is a different cause," a καθ' αὑτὸ συμβεβηκός attribute of an epistemic substance.

> So one definition of definition is what has been said, but another definition is an account that reveals the reason why (λόγος ὁ δηλῶν διὰ τί ἔστιν). So that the first kind of definition signifies, but does not prove, while it is clear that the latter will be as it were a demonstration of the essence, differing from a demonstration in how the terms are placed (ὁ δ' ὕστερος φανερὸν ὅτι ἔσται οἷον ἀπόδειξις τοῦ τί ἐστι, τῇ θέσει διαφέρων τῆς ἀποδείξεως). For, saying why it thunders is different from saying what thunder is. For in the one case you will say "because fire is quenched in the clouds." The other case is "What is thunder? The noise of fire being quenched in the clouds." So that the same account (λόγος) is said in a different way, and while in this way it is a continuous demonstration, in that way it is a definition. (2.10.93b38–94a7)

The demonstration of the existence of something "of which there is a different cause" that was sketched in 2.8 showed how those attributes that are taken as the nominal definition of the definiendum exist because they are logically entailed by predications that do not demand explanation. The scientifically adequate definition of such a kind, one that will reveal the cause, will be the demonstration's major term (the nominal definition of the kind), identified as inherent in its proper sub-

62. According to my reading, in this paragraph Aristotle seems to be considering as accidental any grasp of a thing's existence that is not by virtue of its whole real definition, while at 2.8.93a21 he reserves the word *accidental* for those cases in which one grasps a thing's existence through the sort of nominal definition that does not express any predicates that are present in the thing's real essence. Ackrill ("Aristotle's Theory," 376) writes that, according to this "line of interpretation both passages are speaking of the preconditions for successful investigation. They are both contrasting favourable and unfavourable starting points. But whether the line between favourable and unfavourable is drawn in the same place in both passages is doubtful." On this, see also Bolton, "Essentialism," 535–40; and Sorabji, "Definitions," 217.

ject (the demonstration's minor term) and conjoined with the attribute of being caused by one of the demonstration's middle terms. Aristotle returns to the example of thunder. The nominal definition of thunder asserts that it is "a certain noise." In demonstrating why thunder exists, it is shown that this noise inheres in clouds. This demonstration, whose conclusion is "clouds are such as to have noise in them," has as a middle term "quenching of fire." This demonstration reveals the very same fact as the scientific definition of thunder that can be culled from it: "thunder is the noise of fire being quenched in the clouds." Only the linguistic formulation is different.

Aristotle next identifies the demonstration's minor term, identified as inherent in its major term, as a kind of definition: "Further, a definition of thunder is noise in the clouds, and this is a conclusion of the demonstration of what it is" (2.10.94a7–9).[63] In many cases the definition that expresses the meaning of the term (the first kind of definition distinguished in 2.10) is also an instance of this third kind of definition, the conclusion of a demonstration that reveals the essence of the referent of the term.[64]

---

63. This sentence has been taken to support the interpretation of the demonstration of the definition of a kind K, according to which the conclusion of that demonstration is of the form "a K is such and such." See Ackrill, "Essentialism," 361. Such an interpretation would call into question the interpretation of this demonstration I have sketched here as well as the role in Aristotle's theory of explanation that I have ascribed to it. But I do not see that 2.10.94a7–9 necessitates this alternative interpretation. Aristotle often employs ὅρος to refer to the definiens alone, not the predication of the definiens of the definiendum (cf. *Top.* 1.5.101b38; and the example at *Metaph.* 5.12.1020a4–6). The definiens Aristotle has in mind is "noise in clouds" (ψόφος ἐν νέφεσι). How can such a definiens be taken to be a syllogistic conclusion, which must be a proposition? The conclusion of the demonstration, as I understand it, is "the clouds are such as to have noise in them," or, more simply put, "noise is in clouds" (ἔστι ψόφος ἐν νέφεσι). This Greek sentence could also be translated as "there is noise in clouds," which is the assertion that the complex predicate given in the definiens exists. We can therefore see how Aristotle could be led to calling the definiens of such a definition the conclusion of the demonstration explaining the existence of the kind in question.

64. Ross (*Aristotle's "Prior and Posterior Analytics,"* 635) identifies these two kinds of definition without qualification (cf. Sorabji, "Definitions," 215; and Demoss and Devereux, "Essence," 135–36). Ross is supported by the fact that, in the recapitulation of the kinds of definition given at the close of *APo.* 2.10 (94a11–14), Aristotle lists only three kinds: an indemonstrable expression of essence, the syllogism of the essence (which differs from a definition "in how the terms fall"), and the conclusion of the demonstration of the essence. Absent from this list is a nominal definition as such. But a nominal definition will not be of the third kind if (1) it is of a nonreferring term (on this case, see chap. 3, sec. E), (2) it falsely identifies some real attribute as inherent in a real subject (e.g., "turning away of the Moon" as a nominal definition of the lunar eclipse [cf. 2.8.93b5]), or (3) it has

The last sort of definition Aristotle distinguishes is "an indemonstrable positing (θέσις) of the 'what it is' " (94a9–10). This is the sort of definition with which we are familiar from *Posterior Analytics* 1.2.72a18–24, a first principle by which there is expressed the essence of an epistemic substance.[65]

We see that *Posterior Analytics* 2.10 is wholly consonant with and supportive of my account of the "demonstration of an essence" in 2.8.

F. Themistius on *Posterior Analytics* 2.9

The line of interpretation of 2.8–10 that I have put forward rests on taking the distinction between what has "a different cause" and what does not to be the distinction between the attributes the sciences consider to be inherent in epistemic substances and the epistemic substances themselves. In this I have been anticipated by Ross and, perhaps, by Themistius. I here discuss Themistius' suggestive though sketchy commentary on 2.9.

I provide the Greek text, followed by an English translation of Themistius' paraphrase:

> Ἐκεῖνο μέντοι γε ἀναγκαῖον διορίσασθαι, ὡς οὐκ ἐφ' ἁπάντων τοῦτο συμβαίνει ὧν ζητοῦμεν τοὺς ὁρισμούς, ἀλλὰ τῶν ἄλλοις ὑπαρχόντων, οἷον ἐκλείψεως, βροντῆς· ταῦτα γὰρ τὰ ἀποδεικτά, καὶ ἄλλο ἐπὶ τούτων τὸ μέσον τε καὶ τὸ αἴτιον τοῦ συμπεράσματος καὶ τοῦ πράγματος. ἐπὶ δὲ τῶν καθ' αὑτὰ ὄντω ν καὶ μὴ ἐν ἄλλῳ οὐκετι οἷον τε ἐφ' ἁπάντων τῇ αὐτῇ χρήσασθαι μεθόδῳ· οὐ γὰρ τὸ <τοῦ> εἶναι αἴτιον ἄλλο ἐστὶν ἐφ' ἁπάντων, οἷον τῶν πρώτων καθ' ἑκάστην ἐπιστήμην, ἃ καὶ εἶναι καὶ τί ἐστιν ὑπο-

---

as the definiens some wholly accidental attribute of the definiendum, which cannot be demonstrated to inhere in its subject. See Barnes, *Aristotle's "Posterior Analytics,"* 209; Deslauriers, "Aristotle's Four Types," 20–21.

Another reason why these two kinds of definition cannot be identified point blank is that there is no reason why a nominal definition must identify the subject in which the definiens inheres (although all of Aristotle's examples of nominal definitions do so at least implicitly).

Deslauriers ("Aristotle's Four Types," 1–2, 19–21) suggests that the first kind of definition identified in 2.1 is the nominal definition of only what has "no different cause," but she does not show why this must have been what Aristotle has in mind.

65. Cf. Aquinas, *In Aristotelis*, L.II,l.viii,488.

θέσθαι δεῖ ἢ ἄλλον τρόπον φανερὰ ποιῆσαι, ἐξ ἐπαγωγῆς ἢ πίστεως ἢ ἐμπειρίας, καὶ οὐ δι' ἀποδείξεως τὰ τοιαῦτα· ἀρχαὶ γάρ εἰσιν ἀποδείξεως.[66]

[Nonetheless, the following distinction needs to be made: this kind of demonstration of a definition does not apply in the case of all things of which we seek the definition, but only in the case of those, such as an eclipse and thunder, that belong to other things. For these are the demonstrable things, and in these cases the middle term of the conclusion, i.e., the cause of the thing, is something different from the thing. But in the case of things that are καθ' αὐτά and do not inhere in something else, it is no longer the case that this same method is applicable in all cases; for the cause of existence is not different in every case. For example in the case of things that are primary in respect to each science, it is necessary to hypothesize the existence and the "what it is" or make them clear in some other way (from induction or belief or experience), and such things cannot be made clear through demonstration. For they are principles of demonstration.]

As I have interpreted Aristotle, those things that have no cause different from themselves are those things that are called καθ' αὐτά at *Metaphysics* 5.18.1022a32–35, the things that some science considers to be an ultimate substrate of predication. In identifying these as the beings that are καθ' αὐτά, not predicated of anything else, Themistius is apparently putting forward the same interpretation.

There are two difficulties. First, Themistius' account of the demonstration of the essence is very sketchy. He does not show exactly how the definitions of subjects that are *not* καθ' αὐτά can, in a sense, be demonstrated. Second, Themistius finds in the discussion of the senses of καθ' αὐτό in *Posterior Analytics* 1.4 a sense of καθ' αὐτό according to which the term denotes what has "no different cause." It is not, as one might expect, the sense discussed at 73b5–8, which Themistius takes to refer to accidents "when they are not predicated of other first things, for example, being colored of the surface" (an interpretation, which I find unwarranted). Rather, the fourth sense of καθ' αὐτό delineated in 1.4 (in

66. Themistius, *Analyticorum Posteriorum Paraphrasis*, ed. M. Wallies, *Commentaria in Aristotelem Graeca* 1, no. 5, Prussian Academy Edition (Berlin: George Reimer, 1900), 50, l. 19–28.

73b10–16) is said to be applicable to any term described as follows:

> . . . not predicated of something else but predicated in an absolute sense (ἁπλῶς), because it is predicated neither on account of something else nor of something else; for such are the individuals in respect to essence. In this way "this Callias" is said, as well as some καθ' αὑτά causes: all of those that are neither accidentally predicated of something nor are a cause only by virtue of the occasion (κατὰ σύμπτωμα), for example, if there is lightning while someone is taking a walk. For it is not the case that there was lightning because one was taking a walk, but it was an accident . . .[67]

The examples of καθ' αὑτό and accidental causation are Aristotle's, taken from the passage Themistius is paraphrasing. But it is not at all clear how this text can be taken as offering an account of what is predicated in an absolute sense, not properly predicated on account of something else nor of something else. For example, "dies" *is* predicated on account of the exemplification of another predicate: "is slaughtered"; further, it can be understood only as predicated of "animal." "Dies" cannot be abstracted from its metaphysical substrate for the sake of study as can those epistemic substances that are not truly substantial.

Although Themistius' understanding of the relevant texts is incomplete and defective, he is the first of the extant commentators who perceived the correct line of interpretation of *Posterior Analytics* 2.8-9.

67. Ibid., 11, l. 10–15.

6

On Demonstrating καθ' αὑτὰ συμβεβηκότα Propositions

A. The Importance of *Posterior Analytics* 2.8

According to the interpretation of 2.8–10 that I have presented, Aristotle asserts that for those beings studied by a science that are not epistemic substances, there can be a demonstration that shows how the inherence of this being in an epistemic substance follows from the first principles of the sciences. There must first be a nominal definition, which initially identifies the being in question with some conjunction of attributes; the demonstration proceeds by showing how that conjunction of attributes is predicated of some epistemic substance. By conjoining the definiens of this nominal definition with one of the middle terms of this demonstration, one arrives at a scientifically adequate definition of this being. But this demonstration plays a larger role in Aristotle's philosophy of science than merely being the means by which a few phenomena are explained. Rather, the form of demonstration Aristotle presents will be the mainstay of an Aristotelian science, for it is this kind of demonstration that can do more than make manifest genus-species relations. Because any καθ' αὑτὸ συμβεβηκός predicate has "a cause other than itself," it will be the form of the demonstration of *any* καθ' αὑτὸ συμβεβηκός predication.

It is true that Aristotle has neither argued nor indicated that this demonstration is the *only* way in which a nondefinitional attribute, καθ' αὑτὸ συμβεβηκός of a subject, can be demonstrated to inhere in that subject. It is, however, the best way in which one can meet the difficulties I have outlined in chapter 1, section A concerning the possibility of demonstrating a καθ' αὑτὸ συμβεβηκός predication.

We recall that because every ultimate premise of a demonstration is implicitly definitional, an Aristotelian demonstration appeared to be capable of explaining only trivial propositions. There appeared to be no means by which any term other than one that is implicitly definitional of the demonstration's minor term could appear in the demonstration. Yet the καθ' αὑτὰ συμβεβηκότα predicates, which Aristotle elsewhere *does* take to be demonstrable of a subject of the sciences, are not implicitly demonstrable of their subjects.

But we now see that the sort of demonstration sketched in *Posterior Analytics* 2.8 allows Aristotle's theory to be not bound by these limitations. It is possible to demonstrate the inherence in some epistemic substance S of a kind K that is not implicitly definitional of S, if K can be identified with some conjunction of predicates that *are* implicitly definitional of the subjects of the sciences. Since Aristotle does not present in full the form of the demonstration of the essence of "that of which there is a different cause," the examples Aristotle presents of definienda of nominal definitions, that are demonstrated to exist, are not analyzed to the point where one can see how they are conjunctions of implicitly definitional predicates. But it is not hard to see that such analyses, cumbersome though they may be, would be possible for each of Aristotle's two examples of what has "a different cause": the lunar eclipse and thunder. The phenomenon of the lunar eclipse could be analyzed as a conjunction of predicates that are implicitly definitional of the epistemic substances of astronomy and optics. For the lunar eclipse is identified as what occurs when the Moon, the Sun, and Earth are positioned in such a way as to have light from the Sun blocked from the Moon, causing the Moon to have a certain appearance. Likewise, the phenomenon of thunder could be analyzed as a conjunction of predicates that are implicitly definitional of the epistemic substances of chemistry, acoustics, and meteorology, for (on the account Aristotle here considers) thunder is identified as a sound made by the quenching by water of a certain fire that occurs in the clouds. To demonstrate that each phenomenon exists is to demonstrate that the terms of each conjunction all inhere in some single subject.

As we have seen in chapter 3, the phrase "that of which there is no different cause" signifies for Aristotle an epistemic substance. It follows that "that of which there is a different cause," the sort of being whose definition can in a sense be demonstrated by the sort of demonstration sketched in 2.8 (2.8.93b19, 2.9.93b25–28), is some being that is not an epistemic substance (whose existence is self-explanatory) but has its existence caused by some other being or beings that are to be numbered among the epistemic substances. The causal bonds that are identified in scientific explanation are primarily instances of the "essential cause."[1] "That of which there is a different cause" will be some attribute of an epistemic substance whose existence is logically entailed by that epistemic substance, though it is not implicitly definitional of it. In other words, it will be καθ' αὑτὸ συμβεβηκός of that subject.

B. A Biological Example

Within the *Posterior Analytics* Aristotle presents very few examples of demonstrations other than the examples of the lunar eclipse and thunder expounded in 2.8. Most other examples are mere allusions that reveal nothing of the logical form of the demonstrations that Aristotle has in mind. There is, however, one demonstration of the inherence of a καθ' αὑτὸ συμβεβηκός attribute that is more or less fully presented in the *Posterior Analytics*. This is the botanical demonstration, discussed (in the context of a different inquiry) in *Posterior Analytics* 2.16–17, that a certain flat-leaved plant, such as a vine, is such as to shed its leaves. This demonstration can be shown to have the form of those described in 2.8.

The syllogisms as Aristotle presents them are as follows:

A) (i) Flat-leaved plants are plants such that the sap solidifies at the juncture of the seed,
(ii) any plant such that the sap solidifies at the juncture of the seed sheds its leaves,
so, (iii) flat-leaved plants shed their leaves. (2.17.99a25–29)

B) (i) A vine is a flat-leaved plant,
(ii) flat-leaved plants shed their leaves,
so, (iii) a vine sheds its leaves. (2.16.98b5–10, 2.17.99a24–25)

1. D. Graham (*Aristotle's Two Systems* [Oxford: Clarendon Press, 1987], 159–60) has argued that the Aristotelian notion of an explanation on the basis of some essence predates the analysis of composite substance into matter and form and, hence, cannot be regarded as indicating the "formal" cause of *Phys.* 2.3.194b26–29 (though the identification of form and essence in Aristotle's later metaphysics leads to the identification of the two kinds of cause [see Graham, *Aristotle's Two Systems*, 168, 173–74]). Insofar as all true demonstrations are grounded on definitions, which express essences, all explanatory definitions will reveal an essential cause. In *APo.* 2.11 Aristotle attempts to show that the middle terms of syllogisms can also serve to identify an efficient cause, a final cause, or a cause such that "when these things are so, this necessarily is," but this chapter is notoriously obscure and problematic.

This can be only a partial analysis. It is true that premise i of syllogism B is definitional and premise ii of syllogism B is the conclusion of syllogism A. But, since both premises of syllogism A demand explanation, the whole demonstration as presented cannot impart scientific understanding of its conclusion (cf. 2.8.93a35–37). But, insofar as Aristotle presents these syllogisms as explanatory demonstrations, he takes both of these premises to be demonstrable. How might one demonstrate premise ii of syllogism A?[2]

The subject of this proposition is "plant such that the sap solidifies at the juncture of the seed"; its predicate is "sheds its leaves," which is nondefinitional but, presumably, demonstrable of its subject. "Sheds its leaves" is neither a subject nor a genus nor a differentia of any epistemic substance considered by a science. If the term denoting this attribute is to be employed in a demonstration at all, it can only be by virtue of a presupposed grasp of the meaning of a term, along with a demonstration explaining how it is the case that this meaning is exemplified in reality. There must be a syllogism of the essence of leaf shedding.

The conclusion of the syllogism would be a proposition in which the nominal definition of "sheds its leaves" is predicated of "plant such that the sap solidifies, etc." Let us say that the nominal, pretheoretical definition of "sheds its leaves" is "has its leaves move toward the center of the earth." We can then sketch the following demonstration:

C) (i) The leaves of any plant are primarily made up of earth and water,
(ii) any body that is primarily made up of earth and water has a natural motion toward the center of Earth,
so, (iii) the leaves of any plant have their natural motion toward the center of Earth.

D) (i) Any plant such that the sap solidifies at the juncture of the seed has leaves whose natural motion is unimpeded,
(ii) any plant having leaves whose natural motion is unimpeded

2. R. McKirahan (*Principles and Proofs: Aristotle's Theory of Demonstrative Science* [Princeton: Princeton University Press, 1992], 169) sees the problem, and addresses it by saying that "the congealing of sap" is an element in the "fat definition" of *deciduous* (the term I translate as "shedding leaves"). On McKirahan's conception of "fat definitions" in Aristotle, and its inadequacy in solving the problem of the demonstration of καθ' αὑτὰ συμβεβηκότα predications, see chap. 1, sec. B.

has leaves moving toward the center of Earth,
so, (iii) any plant such that the sap solidifies at the juncture of the seed has leaves moving toward the center of Earth.

E) (i) Any plant such that the sap solidifies at the juncture of the seed has leaves moving toward the center of Earth.
(ii) any plant whose leaves move toward the center of Earth sheds its leaves,
so, (iii) any plant such that the sap solidifies at the juncture of the seed sheds its leaves.

Premise i of C is demonstrable on the basis of the definition of "leaves," which would express its final cause, for presumably one could show that the leaves fulfill their function only if they are made up of water and earth.[3] Premise ii of C follows from the principles of physics (cf. *DC* 4.4.311b4–9). Premise i of D follows from the principles of mechanics, as applied to the structure of certain kinds of plants (of which there is an account in the principles of botany). Premise ii of E is the nominal definition of "sheds its leaves." I will not discuss how premise i of A can be demonstrated, but it should be apparent that this could similarly be accomplished on the basis of the principles of botany and chemistry.

If my account of 2.8–10 is correct, this demonstration, that certain plants shed their leaves, could also serve as a kind of demonstration of the definition of "leaf shedding." The definition arrived at would be "having leaves move toward center of Earth on account of the solidification of the sap at the juncture of the seed." That this is indeed Aristotle's intent is confirmed by his saying that "solidification of the sap" is the definition (λόγος) and "what it is" (τί ἐστί) of "leaf shedding" (2.17.99a25–29).

This, the only full sketch of a demonstration outside of *Posterior Analytics* 2.8, can accordingly be interpreted as having the structure of the kind of demonstration discussed in 2.8. This further supports the view that Aristotle conceived of this demonstration as the kind by which καθ' αὑτὰ συμβεβηκότα predications are to be explained.

3. Cf. *Phys.* 2.9.200a7–30 and *Metaph.* 8.4.1044a27–29, in which Aristotle indicates that in nature, as in art, attaining a certain goal often necessitates that there be a certain kind of matter. For example, there can be no saw if there is no iron (or matter with similar rigidity and hardness).

C. On the Commensurate Universal

The account of the logical form of καθ' αὑτὰ συμβεβηκότα predications that has been presented allows us to solve a long-standing problem in the interpretation of the *Posterior Analytics*.[4] There are indications that Aristotle holds that in a well-formed demonstration, that is, in one in which the major term is explained by the minor term, and not by any genus to which the minor term may belong, the major and minor terms will convert. In other words, if the καθ' αὑτὸ συμβεβηκός conclusion of a well-formed demonstration is "S is P," it is not only the case that it is necessary that any S has P predicated of it; P will be found as a predicate of S alone. The received view is that Aristotle does indeed subscribe to this thesis. But Aristotle's reasons for holding this view have escaped interpreters of the *Posterior Analytics*, to the point that Barnes has argued that Aristotle could not have held such a view at all.[5]

Here I examine the key evidence for Aristotle's holding the doctrine of the commensurate universal. I then show how this doctrine is easily explained on the basis of the results of chapter 6 concerning Aristotle's semantic analysis of καθ' αὑτὰ συμβεβηκότα terms and the logical form of the demonstration of καθ' αὑτὰ συμβεβηκότα predications.

Aristotle explicitly considers the problem of whether the subject and predicate of a demonstrated predication convert in *Posterior Analytics* 2.16–18. These chapters appear to present alternative independent treatments of the problem, one in 2.16 and another in 2.17–18. Each treatment contains arguments both for and against the thesis that there is such conversion. One could take this as evidence that Aristotle's discussions are aporetic throughout.[6] Alternatively, one could go the traditional route and take the arguments for the thesis to be conclusive and endorsed by Aristotle, while taking the arguments against the thesis to be mere qualifications of his position. On this understanding Aristotle holds that a subject and a predicate demonstrated to hold of that subject convert except when the subject belongs to a more general

4. Though he is skeptical about the general line of interpretation taken in the present book, it was Thomas Upton who suggested to me that a strength of the interpretation that I offer is its providing a way to solve the problem of the role of commensurate universals in Aristotle's theory of demonstration.

5. J. Barnes, *Aristotle's "Posterior Analytics," Translated with Notes* (Oxford: Clarendon Press, 1975), 247–48. McKirahan (*Principles and Proofs*, 176) is likewise skeptical that Aristotle holds that all premises (and conclusions) of demonstrations are convertible.

6. Barnes, *Aristotle's "Posterior Analytics,"* 246.

kind, all of whose members have predicated of them the attribute in question (2.16.98b25–31, 2.17.99a8–11, 99a30–b3), or when the predicate in question is not a true unity but is merely called by a single name (2.17.99a7, 11–16).[7]

Throughout Aristotle's positive argument for the convertibility of a subject and a demonstrated predicate rests on the convertibility of the *premises* of the demonstration. This is most fully spelled out at 2.16.98b32–38, in which Aristotle writes:

> Or is this not rather how things stand? If every problem is always universal, is the cause some whole, and is that of which it is the cause a universal? For example, shedding of leaves is determined to some whole (ὅλῳ τινὶ ἀφω ρισμένον), even if there are species of it, and it belongs to these universally, either to plants, or to plants of such a kind. So in these cases the middle and that of which it is a cause must be equal and convert. For example, why do trees shed their leaves? Well, if it is because of the solidification of the sap, then if a tree sheds its leaves, there must be solidification, and if there is solidification, not of anything whatsoever but of a tree, there must be shedding.

We have already considered the example Aristotle presents in the first passage: the biological demonstration that vines shed their leaves. In section B our interest was the question of the logical form of such a demonstration, and we attended to Aristotle's account because it conforms to the account of the demonstration of καθ' αὑτὰ συμβεβηκότα predications that we have discerned in 2.8–10. But in context this example has another role to play. It is meant to clarify to what extent and for what reasons the subject and the predicate of a demonstrated proposition must convert.

Aristotle's conclusion that there is such a conversion is conditional on it being the case that "every problem is always universal." For he claims that, if that is so, "the cause is some whole, and that of which it is the cause is a universal." As an example of this, he points to the conversion of the middle and major terms of the demonstration that a tree

7. The best detailed account of *APo.* 2.16–18 is that of W.D. Ross (*Aristotle's "Prior and Posterior Analytics"* [Oxford: Clarendon Press, 1949], 666–73), who understands Aristotle's argument in this way.

sheds its leaves. For shedding leaves and solidification of the sap convert.

Three problems present themselves in our interpretation of this passage. First, has Aristotle not already considered the case of a demonstration that *vines* shed their leaves? If both vines and trees shed their leaves, is this not a counterexample to the thesis of the convertibility of subjects and demonstrated predicates? Second, what sense do we make of the talk of "universality" that is meant to account for the convertibility of the middle and major terms? A third and related question is why Aristotle takes it to be unproblematic that the middle and major terms of his example will convert. If we grant that solidification of the sap serves as an adequate scientific cause of shedding, it follows that everything whose sap solidifies (at the connection of the seed [2.17.99a28–29]) will shed its leaves. But why can there not be cases of the shedding of leaves that have a different cause—for example, torrential winds, or even some other natural biological process?

Aristotle explicitly addresses the first question. A tree does not shed its leaves insofar as it is a tree nor does a vine shed leaves insofar as it is a vine. Rather, there is a genus of plant to which they both belong, that of being a broad-leaved plant, and it is this genus that is responsible for the shedding (2.16.98b5–10, 2.17.99a23–25). In this case *shedding* holds of every instance of the subject, and only of instances of that subject; the terms convert. Aristotle sometimes calls such a convertible term a *primary universal* (πρῶτον καθόλου) (1.4.73b32–74a3, 74a36–74b4; 1.5.74a10–13; 2.17.99a33–35), for the most generic subject to which the predicate belongs universally is the first (πρῶτον) of a series to which it belongs (1.4.73b32–39).[8] Aristotle's explicit recognition of the need to demonstrate that a term is predicated of that to which it belongs primarily leads us to conclude that he is speaking loosely when he declares that the shedding that is of interest in the demonstration that trees shed is the shedding of *trees*, not the shedding of anything at all (2.16.98b38). He should have said that what is of interest is not the

8. The discussion by B. Inwood ("A Note on Commensurate Universals in the *Posterior Analytics*," *Phronesis* 24 [1979], 321–22) of these passages is very helpful. He errs in maintaining that the definition of πρῶτον at 2.13.99a33–35 entails that the πρῶτον universal is predicated of its subject καθ' αὑτό. If I enjoy all green fruits, and only green fruits, "green fruit" is a πρῶτον universal in respect to "fruits I enjoy" but does not hold καθ' αὑτό of "fruits I enjoy." Commensurate universals are both πρῶτον and καθ' αὑτό (insofar as they are objects of scientific explanation).

shedding of anything at all but, rather, the shedding of *broad-leaved plants.*

But does this not beg the question? If we declare that the predicate that is to be explained must be considered only as inherent in a given subject, will it not be trivially true that the subject and the predicate convert? If not, what justification is there for this restriction on how one is to understand the attribute under investigation?

We are led to the other two questions. First, how could the fact that both the attribute under consideration and its cause are "universal" ensure that these are convertible? If all B's are C's but not all C's are B's, we are taking both B and C to be universals, insofar as both C and B are considered as predicated of a multiplicity of particulars (*DI* 7.17a39–40), but these two terms do not convert.

Aristotle is clearly using the term *universal* in a special sense. We need not look outside the *Posterior Analytics* to find this sense delineated. For at 1.4.73b26-27 Aristotle writes, "I mean by 'universal' what belongs to every instance of its subject and in respect to the subject and insofar as the subject is what it is (καθόλου δὲ λέγω ὃ ἂν κατὰ παντός τε ὑπάρχῃ καὶ καθ' αὑτὸ καὶ ᾗ αὐτό)." To say that the term is predicated of every instance of the subject is to say that it is what Aristotle usually calls a universal and that it is predicated universally of its subject. To say that it is predicated in respect to its subject is to say that between the subject and predicate there is a causal bond of the sort necessary for the predication to be an object of scientific understanding (1.4.73b16–18).[9] Because Aristotle at 73b28–29 indicates that he is using "in respect to itself" in a special sense, according to which the phrase is synonymous with "insofar as it is itself," a universal predication is not sufficiently defined as one in which the predicate hold of every instance of the subject and in which there is a causal connection between subject and predicate. One must add that in a universal predication the attribute is

9. It is clear that all predications that are objects of scientific understanding are καθ' αὑτά in some sense. Definitional predications are καθ' αὑτά insofar as the predicate is present in the essence of the subject (*APo.* 1.4.34–36; *Metaph.* 5.18.1022a27–29), and non-definitional demonstrated predications are καθ' αὑτά in some sense, for they are called the καθ' αὑτὰ συμβεβηκότα. These predications seem to be called καθ' αὑτά insofar as the predicate holds κατά the subject, and the preposition κατά has the philosophic function of indicating a cause (*Metaph.* 5.18.1022a14–22). But it is very unclear how the καθ' αὑτὰ συμβεβηκότα predications have a place in either of Aristotle's two classifications of καθ' αὑτά predications (*APo.* 1.4.73a34–b16 and *Metaph.* 5.18.1022a24–36). To my mind there has not yet been an adequate account of this issue.

caused only by that subject (i.e., the subject and the attribute convert) and that the subject has been identified with the appropriate degree of specificity.[10]

We have shown only that Aristotle's terminology in *Posterior Analytics* 2.16 is consistent with that of 1.4. Left unresolved is the third and main issue: why we should hold that a demonstrated attribute has only one cause.

Aristotle's way of putting his point here is to say that the attribute is "determined (ἀφωρισμένον) to some whole." Barnes and Inwood take this to be a brute assertion of convertibility. Finding such an assertion to be without warrant, Barnes understands the whole argument as aporetic and denies that Aristotle could have endorsed as his considered judgment the thesis that an attribute and its cause necessarily convert.[11] Inwood takes Aristotle at his word here but expresses puzzlement over why Aristotle should have held that the convertibility of a καθ' αὑτὸ συμβεβηκός attribute and its cause is self-explanatory.[12]

We should note that ἀφωρισμένον has as its root the verb ὁρίζειν, which not only has the sense of "to determine" but is also Aristotle's main term for "to define." Could it not be the case that Aristotle is here indicating that a single cause of the demonstrated attribute is indicated in the definition of that attribute? An alternative translation of the phrase in question could be "shedding of leaves is defined in regard to some whole."

We have seen that Aristotle makes precisely this point in 2.8–10. The scientific definition of a καθ' αὑτό attribute is a conjunction of an adequate nominal definition of that attribute and its cause, a middle term on the basis of which the existence of that attribute can be explained. It follows that, within a scientific definition of such an attribute, a single cause will be indicated. Consequently, any instance of that attribute necessarily has that cause.

But will it not sometimes be the case that ordinary language employs the same term in respect to an attribute with two or more different independent causes? To return to the example presented earlier: Could we not say that of a tree whose leaves were stripped by torrential winds that it has shed its leaves? This would conform to everyday usage but not to scientific usage, for within scientific discourse one is bound to

10. On this passage see esp. McKirahan, *Principles and Proofs*, 95–98.

11. Barnes, *Aristotle's "Posterior Analytics,"* 242–43.

12. Inwood, "Notes," 324–25.

the scientific definition of shedding. To take another example: let us take it as given that a certain noise in the clouds, called thunder, is caused by quenching of fire in the clouds. But suppose that another sound (which the ear cannot discriminate from quenching of fire in the clouds) were caused by colliding masses of air. Prior to any meteorological inquiry, people would use the term *thunder* in respect to each. Only science would show that the predicates expressed in the nominal definition of this term (noise in the clouds) are exemplified for two different reasons. If I am correct, Aristotle's response would be that in these cases *shedding* and *thunder* are homonymous, though no one could have been aware of this prior to the scientific inquiry.[13] This does not seem an unreasonable response. Contemporary essentialists such as Putnam and Kripke have made much the same point in maintaining that terms that denote natural kinds are "indexical," or "rigid designators," referring to natural kinds whose members are those specimens having certain features that scientific authorities eventually determine to be characteristic of that kind.[14]

We see that the conversion of a καθ' αὑτὸ συμβεβηκός attribute (as scientifically understood) and its cause follows from the argument of *Posterior Analytics* 2.8–10, as it has been interpreted here. If a demonstration has more than one middle term, each middle term will convert with that before it, until we reach a middle term that is definitional of its subject. If the demonstration is well formed, that middle term will not be the genus of the subject. (Otherwise, it will be that genus, and not a subject that falls under that genus, that is the true cause of the inherence of the attribute in question.) It must therefore be the differentia of the subject. Throughout his career Aristotle maintained that a differentia converts with its subject (cf. *Cat.* 3.1b16–24; *Metaph.* 7.12). Accordingly, every demonstrated attribute, as scientifically understood, converts with its subject.

13. I grant that within *APo.* Aristotle does not envisage the possibility of this sort of homonymy, that of two terms with the same nominal definitions but different scientific definitions. The example of homonymy he does consider is that of *similarity* in respect to colors and figures (2.17.99a12–15); it is not clear that in each case there would be the same nominal definition of *similarity*.

14. See Kripke, *Naming and Necessity*, rev. and enl. ed. (Oxford: Blackwell, 1980); Putnam, *Mind, Language, and Reality: Philosophical Papers, Volume 2* (Cambridge: Cambridge University Press, 1975), 196–271.

D. A Problem: *Posterior Analytics* 1.9 Belied

There is a problem with my account of the demonstration of καθ' αὑτὰ συμβεβηκότα predication. It can be shown that any such demonstration must rest on the principles of more than one science, thereby violating the demand Aristotle lays down in *Posterior Analytics* 1.7 and 1.9 that a demonstration of a predication must rest on principles of only that science that is concerned with the predication's subject. I here argue that it is indeed the case that a demonstration of a καθ' αὑτὸ συμβεβηκός predication must rest on the principles of more than one science. I argue that for this reason Aristotle's demand that the principles of a demonstration be concerned with only a single kind is not consistent with the rest of his theory of demonstration. I show that Aristotle was at least implicitly aware that this is so.

As we have seen, a καθ' αὑτὸ συμβεβηκός predicate is identified with a certain conjunction of predicates through a nominal definition. Since the whole conjunction is not definitional in regard to a subject S of which it is predicated, at least one of the predicates in terms of which the καθ' αὑτὸ συμβεβηκός predicate is defined is not definitional in regard to S. It follows that a demonstration of that conjunction of predicates that is identified as a καθ' αὑτὸ συμβεβηκός predicate must either be impossible or rest on a premise that is itself καθ' αὑτὸ συμβεβηκός in relation to the subject. But such a premise must itself be demonstrated if it is to have the status of an object of scientific understanding. We are threatened by a regress, but the demonstrations of καθ' αὑτὰ συμβεβηκότα predications that we have examined shows how this can be avoided. The demonstration of the shedding of leaves had as its conclusion "leaves fall"; the demonstration of the existence of thunder had as its conclusion "clouds make noise." "To fall" and "to make noise" are καθ' αὑτὰ συμβεβηκότα in respect to "leaf" and "cloud," respectively. But, for Aristotle, "falls" *is* definitional for some subjects considered by chemistry (water and earth) and "makes noise" is definitional for some subject (perhaps air) considered by acoustics. We can regard a leaf merely as a certain body and a cloud merely as a noise-making body, not on the basis of the principles of botany and meteorology, the sciences that posit as objects of study the plants and the cloud, respectively, but, rather, on the basis of the principles of physics and acoustics. Were it otherwise, the falling of objects and the hissing of quenched fire would not be intelligible, as they must be if the falling of leaves and the noise of thunder are taken to be objects of scientific explanation.

So, the very same predication that is definitional from the perspective of one science can be καθ' αὑτὸ συμβεβηκός from the perspective of another. (Here by *predication* I mean the exemplification of some predicate in some subject.) The same fact will be expressed in two different ways by two different sciences, for each science will refer to the subject of that predication by means of a different description. For a subject can be referred to by its substantial nature but also by means of some quantitative feature or by some feature of the matter of which it is composed.

A demonstration of a καθ' αὑτὸ συμβεβηκός predication will be possible because that being that is the subject of this predication can be referred to in different ways by both its minor term and a middle term. Each of these terms will denote an epistemic substance for a different science. In one premise of such a demonstration, a kind that is an epistemic substance for one science is predicated of a kind that is an epistemic substance for another—for example, "all plants are (primarily) water and earth" and "clouds are (sometimes) water and fire." Although it is perhaps the case that these particular premises are demonstrable, in order to avoid a regress some such premise will need to be indemonstrable. It should be noted that it is not clear how this can be the case, given the classifications of the principles of the sciences that are presented in the *Posterior Analytics*. Perhaps this was one of the motivations behind the position presented in later writings of Aristotle, that a definition of a composite substance must reveal the material component of that substance as well as the form (*Metaph.* 6.1025b30–1026a6, 7.7.1033a1–5, 8.2.1043a2–28; *De An.* 1.1.403a25–b19). The formal part of an essence of a thing expresses its substantial species, while the material part of the essence expresses the persistent matter, whose actual attributes (such as weight and chemical dispositions) persist. If the definition of a substance expresses both the formal and material parts of the essence, we can sometimes employ such a definition to explain the predication of the middle term of a καθ' αὑτὸ συμβεβηκός demonstration of its minor term. For example, if the definition of a deciduous plant expresses the matter of that plant as well as its form, it can be employed to justify and explain the fact that some theoretical primitive of chemistry belongs to some theoretical primitive of biology.

But in 1.7.75a38–b20 and 1.9.75b37–76a9, 29–30, Aristotle argued that, since each science concerns a certain genus, demonstrations concerning subjects falling under some genus must rest only on principles of that science that deals with that genus.[15] This is plainly inconsistent with the very demonstrations of the existence of καθ' αὑτὰ συμβεβηκότα attributes that Aristotle himself sketches. We must therefore regretfully conclude that Aristotle is not entirely consistent. Such a conclusion is not to be made lightly. But in my estimation more is to be gained in the consistency and power of Aristotle's account of scientific explanation as a whole if we accept the interpretation for which I have been arguing, though it renders Aristotle guilty of inconsistency on this point.

I therefore maintain that the requirement that any demonstration rest only on the principles concerning the genus of the minor term be revised. The requirement that demonstrations be "homogenous" needs to be replaced by a requirement that the principles of demonstration must be proper to the sciences considering the *minor* term of the demonstration. For, if that term is defined as a complex of primitive predicates, the demonstration will rest on the principles of the sciences that consider *those* predicates either as subjects or definitional of their subjects. For example, "healing more slowly" could be defined as "torn flesh growing such that the ratio of the area of new flesh to time is smaller," which presumably can be analyzed as a complex predicate defined in terms of the primitives of physiology (which is concerned with flesh as such), geometry (which is concerned with areas as such), and "universal mathematics" (which is concerned with ratios as such [cf. *Metaph.* 13.3.1077b17–20]).

But such a revision is not wholly alien to Aristotle's thinking. In *Pos-*

15. At 1.7.75b14–17, 1.9.76a9–15, and 1.13.78b34–36 Aristotle qualifies this conclusion, noting that there are some sciences that are subordinate to others, as optics is subordinate to geometry and harmonics is subordinate to arithmetic. All of the καθ' αὑτά attributes of a kind considered by a subordinate science are rendered intelligible only by considering them as features of a kind considered by the superordinate science; the subordinate science is the superordinate science "applied." (See T. Heath, *Mathematics in Aristotle* [Oxford: Clarendon Press, 1949], 58–61; Ross, *Aristotle's "Prior and Posterior Analytics,"* 554–55.) This is not the case for the demonstrations of καθ' αὑτὰ συμβεβηκότα predications considered here; not every meteorological fact, for example, is to be explained on the basis of the principles of acoustics. Nonetheless, a comment Aristotle makes in regard to superordinate and subordinate sciences also holds in regard to the principles of a καθ' αὑτὸ συμβεβηκός predication: they deal with genera that are the same, not in an absolute sense (ἁπλῶς) but in some respect (πῇ) (1.7.75b8–9).

terior Analytics 1.13 Aristotle concludes a discussion of the way in which a demonstration τοῦ διότι of a truth considered by one science may rest on the principles of another science with the following passage:

> And many of the sciences that are not such that one is subordinate to the other are related in this way; for example, medicine is so related to geometry. For to know *that* circular wounds heal more slowly belongs to the doctor, but to know the reason why belongs to the geometrician. (1.13.79a13–16)

This example, too, conflicts with the requirement that a demonstration concerning some subject rest on the principles of only that science studying the subject as such.[16] Perhaps *Posterior Analytics* 1.7 and 1.9 belong to earlier strata of Aristotle's thought than do 1.13 and 2.1–10; perhaps Aristotle never saw that this tension is present in his theory of scientific explanation. Whatever the explanation, this passage furnishes more evidence that Aristotle would have endorsed my account of demonstrations of καθ' αὑτὰ συμβεβηκότα predications. For, on the example that Aristotle himself presents, the explanation of the fact that circular wounds heal more slowly than other wounds (however we are to fill in the details) will rest on considering the same thing (the healing flesh) from the standpoint of biology (as part of a substance that has a nutritive soul and is hence able to heal itself) and from the standpoint of geometry (as having a surface whose edges form a circle). The same torn piece of flesh will need to be regarded as an epistemic substance of two different sciences, just as will every subject of a καθ' αὑτὸ συμβεβηκός predication, on the account I have presented.

E. Concluding Remarks

I have argued in this study that the demonstration revealing an essence sketched by Aristotle in *Posterior Analytics* 2.8–10 has a large and important role to play in Aristotle's account of scientific explanation, for it is by means of such a demonstration that one explains a predication that is not implicitly definitional of its subject. Such demonstrations will be the primary vehicles of scientific explanation. One arrives

16. See Barnes, *Aristotle's "Posterior Analytics,"* 129, 153.

at the correct explanation by beginning with a vague conception of the phenomenon to be explained. This will be expressed in a nominal definition of the term by which one refers to this phenomenon, for a nominal definition is a convention that determines in a preliminary way the necessary and sufficient conditions for being the referent of the term. If the phenomenon in question is not part of the essence of an epistemic substance (a theoretical primitive of one of the sciences), it will be explained by showing how it is rendered necessary by two or more epistemic substances having the essences that they have. One is thereby able to present and theoretically justify a scientific definition of the phenomenon in question. This will be the conjunction of the nominal definition with which one begins, on the one hand, and the causal account of this phenomenon that has been ascertained through scientific inquiry, on the other.

In this sort of explanation a phenomenon that is considered a unity from an everyday point of view[17] is in fact *identified* with a complex of scientific primitives, which, taken together, account for those characteristics of the phenomenon by which it is recognized. So, a demonstration of a καθ' αὑτὸ συμβεβηκός predication is in effect a *reductive* explanation of the phenomenon to which the predicate refers.[18] If we take *Pos-*

17. Consider how the ancient Greeks often deified a regularly occurring meteorological or astronomical event, such as "Day," "Night," and "the Winds" in Hesiod's *Theogony*. Mythological accounts often posited such events as primitive and independently subsistent. When the pre-Socratic philosophers offered alternative accounts of such events, showing how they are caused by more basic entities, they were rightly thought to have challenged the "everyday" account offered by the myths.

18. If it seems odd to consider Aristotle to hold that scientific explanation is a matter of reduction, this is because in recent decades much of the discussion on reduction has focused on the possibility of reducing consciousness and the sensible characteristics of things to the theoretical primitives of physics. Aristotle would, of course, deny that consciousness and simple sensible qualities are reducible to more fundamental primitives. On his account intellect, sensible qualities, and the sensitive soul itself would be among the theoretical primitives. My point is that *in those cases in which he holds that scientific explanation is possible* (e.g., thunder, eclipses, and the shedding of leaves), Aristotle holds that these explanations have the logical form of a reductive explanation. In this respect Aristotle's position could not be that of an emergence theorist, who holds that an emergent property can be correlated to a physical property so that anything possessing it also has the emergent property but denies that the emergent property as such can be identified with or rendered intelligible on the basis of the physical property with which it is correlated. Aristotle holds that the theoretical primitives are the basic causes of things and that causal bonds can be revealed and rendered intelligible through scientific explanation. If A is a property emergent on some physical state B, science can determine that A always follows from B. At the same time, one cannot render intelligible the existence and representative characteristics of A, on the basis of B. So, the existence of emergent

terior Analytics 2.8–10 to be an account of a reductive explanation, we find that it is in two respects parallel to the account of reductive explanation that has been advanced by W. Sellars.

First, an explanation of an attribute or phenomenon as a complex of theoretical primitives has ontological implications for both Aristotle and Sellars. For both are scientific realists, taking the theories to tell about the world and not to be mere instruments allowing one to predict future events.[19]

Second, Sellars, like Aristotle, takes *definitions* to be the means by which one expresses the identification of some phenomenon (considered to be a primitive within an everyday framework) with a complex of theoretical primitives. As for Aristotle, such definitions are not merely expressions of the meanings of terms but assertions of what things are.[20]

Although Aristotle may have been the first to articulate the structure of this sort of reductive explanation, such explanations did not begin with him. The atomists (and Plato, in the *Timaeus*) offered explanations of higher-level phenomena on the basis of the natures of microscopic theoretical primitives. The fact that Aristotle did recognize and, in cer-

---

properties would call into question scientific understanding as it is conceived by Aristotle.

This relation between A and B amounts to what J. Kim ("Concepts of Supervenience," *Philosophy and Phenomenological Research* 45 [1984]: 157–63) calls "weak supervenience." "Strong supervenience" entails causal bonds that necessitate the correlation between the two properties. As Kim argues, if we say that C is strongly supervenient on D, we are committed to the view that any instance of C could in principle be reduced to D, though it is quite possible that we are not in the epistemological position to effect the reduction (163–76).

19. For a representative instrumentalist account of reductive explanation, see E. Nagel, *The Structure of Science: Problems in the Logic of Scientific Explanation* (1961; reprint, Indianapolis: Hackett, 1979), 336–97. Opposing such an account, W. Sellars (*Philosophical Perspectives: Metaphysics and Epistemology* [Springfield, Ill.: C.C. Thomas, 1967], 155) writes that an explanatory theory is not a mere "syntactic machine which provides an external systemization of empirical statements."

20. See Sellars (*Philosophical Perspectives*, 153). Again, this view runs against some mainstream accounts of scientific explanation that were current in the mid-twentieth century. See, for example, Nagel, *Structure of Science*, 357–58, 366; and C. Hempel, "Reduction: Ontological and Linguistic Facets," in *Philosophy, Science, and Method: Essays in Honor of Ernest Nagel*, ed. S. Morgenbesser, P. Suppes, and M. White [New York: St. Martin's Press, 1969], 179–90. This follows from the account of scientific explanation held by these thinkers. If science has no role to play in expressing *how things are*, it will not involve the identification of real essences. The only definitions will be expressions of the meanings of terms, or nominal definitions, and the task of science is neither the description of language as currently employed nor the reformation of linguistic usage.

tain cases, endorse this kind of explanation has not often been recognized, most likely because of the prevalent misinterpretations of the chapters in which Aristotle's discussion of it is presented. The natural philosophers of the seventeenth century who advocated striving for such explanations often contrasted them with explanations in the Aristotelian mode, in which every problematic attribute of a thing is accounted for by positing some distinct faculty or essential attribute within the thing's nature.

This is a misunderstanding of Aristotle's views on explanation. We have seen that in *Posterior Analytics* 2.8–9 Aristotle recognizes a sharp distinction between those cases in which reductive explanation is appropriate and those cases in which it is not. He holds that there are terms in our language that denote things with primitive essences and adamantly insists that reductive explanation of the existence or definitional attributes of these things is impossible. As Aristotle elsewhere makes clear (*Metaph.* 9.10.1051b30–33), if one knows what such things are at all, it is by virtue of νόησις, a kind of simple grasp of things as they really are; the notion of reducing them to more basic entities is fundamentally wrongheaded. But there are terms in ordinary language that denote phenomena, events, and attributes that, on account of their regularity or common features, may appear to constitute a simple unity with a real essence but can in fact be reduced to the attributes of more fundamental entities. Such beings are the καθ' αὑτὰ συμβεβηκότα; it is of them that Aristotle considers reductive explanation appropriate.

Aristotle is sometimes taken to have made the extravagant claim, in *Posterior Analytics* 2.19, *Nicomachean Ethics* 6.6, and elsewhere, that intellect (νοῦς) allows us to grasp in an infallible way the essence of everything (substance, attribute, or event) whose attributes admit explanation. If my thesis is correct, Aristotle can be seen as making a more modest claim. In my view (for which I shall not here argue) Aristotle does indeed hold that there are substances and attributes whose essences can be immediately grasped by νοῦς. But he also recognizes that there are regular features of the world that can be rendered intelligible but at the same time are not such as to have their essences immediately grasped by νοῦς. For these aspects of the world, scientific understanding must come about by some other means; presumably, the generation and testing of hypotheses involved in dialectical inquiry will have a large role to play.[21]

21. I leave as an open question whether Aristotle thinks it possible for the theories of a whole science (e.g., meteorology) to be reduced to those of another science (e.g.,

Aristotle, above all, epitomizes the ancient Greek equation of what is, what is intelligible, and what is expressed in language. Nearly every passage in the Aristotelian corpus bears witness to the belief that there can be isomorphism among being, thought, and language. This is seen first and foremost in the *Posterior Analytics*. There Aristotle most clearly maintains that basic beings are rendered intelligible through λόγοι that are the first principles of the sciences, the basic definitions. There he also asserts that causal linkages can be revealed through those λόγοι that are demonstrations, whose logical entailments are reflective of the causal linkages among facts. The present study has shown that in *Posterior Analytics* 2.8–10, Aristotle shows how he takes this to be possible. Beginning with λόγοι that identify the characteristics by which we recognize things, we determine how these characteristics are caused by basic beings, whose essences are expressed in basic definitions. These causal linkages are expressed in a demonstration of a καθ' αὑτὸ συμβεβηκός demonstration. For such complex facts, λόγοι, and explanations as well for simple beings, definitions, and concepts, "the things in speech are symbols of things occurring in the mind" (*DI* 1.16a3–4).

physics). If so, the reduced science would reveal itself to be not a genuine science at all, since its genus is shared by another. In such a case, the subject matter of the first science would need to be reduced to that of the second through the sort of demonstration Aristotle discusses in *APo.* 22.8–10.

Works Cited

Ackrill, J.R. *Aristotle's "Categories" and "De Interpretatione," Translated with Notes*. Oxford: Clarendon Press, 1963.

———. "Aristotle's Theory of Definition: Some Questions on *Posterior Analytics* II 8–10." In *Aristotle on Science: Proceedings of the Eighth Symposium Aristotelicum*, edited by E. Berti, 359–84. Padua: Editrice Antenore, 1981.

Aquinas, T. *In Aristotelis Libros Peri Hermineias et Posteriorum Analyticorum Expositio*, edited by R.M. Spiazzi. Turin, It.: Marietti Editori, 1955.

Bäck, A. *On Reduplication*. Leiden, Neth.: E.J. Brill, forthcoming.

Balme, D. *Aristotle's "De Partibus Animalium" I and "De Generatione Animalium" I*. Oxford: Clarendon Press, 1972.

———. "Aristotle's Use of Division and Differentiae." In *Philosophical Issues in Aristotle's Biology*, edited by A. Gotthelf and J. Lennox, 69–89. Cambridge: Cambridge University Press, 1987.

Barnes, J. "Aristotle's Theory of Demonstration." *Phronesis* 14 (1969): 123–52. Reprinted in *Articles on Aristotle*, vol. 1: *Science*, edited by J. Barnes, M. Schofield, and R. Sorabji, 65–87. London: Duckworth, 1975.

———. *Aristotle's "Posterior Analtyics," Translated with a Commentary*, 2d ed. Oxford: Clarendon, 1994.

———. *Aristotle's "Posterior Analytics," Translated with Notes*. Oxford: Clarendon Press, 1975.

———. "Proof and the Syllogism." In *Aristotle on Science: Proceedings of the Eighth Symposium Aristotelicum*, edited by E. Berti, 17–59. Padua: Editrice Antenore, 1981.

Bolton, R. "Essentialism and Semantic Theory in Aristotle's *Posterior Analytics* II.7–10." *Philosophical Review* 85 (1976): 514–44.

———. "Definition and Scientific Method in Aristotle's *Posterior Analytics* and *Generation of Animals*." In *Philosophical Issues in Aristotle's Biology*, edited by A. Gotthelf and J. Lennox, 120–66. Cambridge: Cambridge University Press, 1987.

Bonitz, H. *Index Aristotelicum*. Revised by O. Gigon. Berlin: Groyten, 1961.

Burnyeat, M.F. "Aristotle on Understanding Knowledge." In *Aristotle on Science: Proceedings of the Eighth Symposium Aristotelicum*, edited by E. Berti, 97–139. Padua: Editrice Antenore, 1981.

———. ed. *Notes on Book Zeta of Aristotle's "Metaphysics."* Study Aids Monograph No. 1. Oxford: Subfaculty of Philosophy, 1979.

Burnyeat, M.F., and J. Barnes. "Socrates and the Jury: Paradoxes in Plato's Distinction between Knowledge and True Belief." *Aristotelian Society* supp. 54 (1980): 173–206.

Demoss, D., and D. Devereux, "Essence, Existence, and Nominal Definition in Aristotle's *Posterior Analytics* II 8–10." *Phronesis* 33 (1988): 133–54.

De Rijk, L.M. *The Place of the Categories of Being in Aristotle's Philosophy*. Assen: Van Gorcum, 1952.

Deslauriers, M. "Aristotle's Four Types of Definition." *Apeiron* 23 (1990): 1–26.

Ferejohn, M. "Definition and the Two Stages of Aristotelian Demonstration." *Review of Metaphysics* 36 (1982): 375–95.

———. "Meno's Paradox and *De Re* Knowledge in Aristotle's Theory of Demonstration." *History of Philosophy Quarterly* 5 (1988): 99–117.

———. *The Origins of Aristotelian Science*. New Haven: Yale University Press, 1991.

Frede, D. "Comment on Hintikka's Paper 'On the Ingredients of an Aristotelian Science.'" *Synthese* 28 (1974): 79–89.

Frede, M. "Substance in Aristotle's Metaphysics." *Essays in Ancient Philosophy*, 72–80. Minneapolis: University of Minnesota Press, 1987.

Freeland, C. "Scientific Method in Aristotle's Meteorology." Photocopy, 1986.

Goldin, O. "The Role of the Demonstration That Explains an Essence in Aristotle's Theory of Explanation." Ph.D. diss. University of Texas at Austin, 1987.

Gomez-Lobo, A. "Aristotle's Hypotheses and the Euclidean Postulates." *Review of Metaphysics* 30 (1977): 430–39.

———. "The So-Called Question of Existence in Aristotle, *An. Post.* 2.1–2." *Review of Metaphysics* 34 (1980): 71–90.

Gotthelf, A. "First Principles in Aristotle's *Parts of Animals*." In *Philosophical Issues in Aristotle's Biology*, edited by A. Gotthelf and J. Lennox, 167–98. Cambridge: Cambridge University Press, 1987.

Graham, D. *Aristotle's Two Systems*. Oxford: Clarendon Press, 1987.

Halper, E. *One and Many in Aristotle's Metaphysics*. Columbus: Ohio State University Press, 1989.

Hamelin, O. *Le système d'Aristote*. Paris: F. Alcan, 1920.

Heath, T. *Mathematics in Aristotle*. Oxford: Clarendon Press, 1949.

———. *Euclid's "Elements," Translated with Introduction and Commentary*. Vol. 1. New York: Dover, 1956.

Hempel, C. "Reduction: Ontological and Linguistic Facets." In *Philosophy, Science, and Method: Essays in Honor of Ernest Nagel*, edited by S. Morgenbesser, P. Suppes, and M. White, 179–99. New York: St. Martin's Press, 1969.

Hintikka, J. "On the Ingredients of an Aristotelian Science." *Nous* 6 (1972): 55–69.

———. "Reply to Dorothea Frede." *Synthese* 28 (1974): 91–96.

———. "The Varieties of Being in Aristotle." In *The Logic of Being*, edited by S. Knuuttila and J. Hintikka, 81–114. Dordrecht, Neth.: D. Reidel, 1986.

Hussey, E. *Aristotle's "Physics," books III and IV, translated with Notes*. Oxford: Clarendon Press, 1983.

Inwood, B. "A Note on Commensurate Universals in the *Posterior Analytics*." *Phronesis* 24 (1979): 320–29.

Jacobs, W. "Aristotle on Nonreferring Subjects." *Phronesis* 24 (1979): 282–300.

Kahn, C. "The Greek Verb 'Be' and the Concept of Being." *Foundations of Language* 2 (1966): 245–65.

———. *The Verb "Be" in Ancient Greek*. Dordrecht, Neth.: D. Reidel, 1973.

Kim, J. "Concepts of Supervenience." *Philosophy and Phenomenological Research* 45 (1984): 153–76.

Klein, J. *Greek Mathematical Thought and the Origin of Algebra*. Translated by E. Brann. Cambridge, Mass.: MIT Press, 1968.

Kripke, S. *Naming and Necessity*, rev. and enl. ed. Oxford: Blackwell, 1980.

Kullman, W. "Die Funktion der mathematischen Beispiele in Aristoteles' *Analytica Posteriora*." In *Aristotle on Science: Proceedings of the Eighth Symposium Aristotelicum*, edited by E. Berti, 245–70. Padua: Editrice Antenore, 1981.

Landor, B. "Definitions and Hypotheses in *Posterior Analytics* 72a19–25 and 76b35–72a4." *Phronesis* 26 (1981): 308–18.

———. "Aristotle on Demonstrating Essence." *Apeiron* 19 (1985): 116–32.

Lee, H.D.P. "Geometrical Method and Aristotle's Account of First Principles." *Classical Quarterly* 29 (1935): 113–24.

Lennox, J.G. "Divide and Explain: The *Posterior Analytics* in Practice." In *Philosophical Issues in Aristotle's Biology*, edited by A. Gotthelf and J. Lennox, 90–119. Cambridge: Cambridge University Press, 1987.

Liddell, H.G., R. Scott, and H.S. Jones. *A Greek-English Lexicon*. Oxford: Clarendon Press, 1940.

Lloyd, A.C. "Necessity and Essence in the *Posterior Analytics*." In *Aristotle on Science: Proceedings of the Eighth Symposium Aristotelicum*, edited by E. Berti, 157–71. Padua: Editrice Antenore, 1981.

Lukasiewicz, J. *Aristotle's Syllogistic from the Standpoint of Modern Formal Logic*, 2d ed. Oxford: Clarendon Press, 1957.

Mansion, S. *Le jugement d'existence chez Aristote*, 2d ed. Louvain: Publications Universitaires de Louvain, 1976.

McKirahan, R. *Principles and Proofs: Aristotle's Theory of Demonstrative Science*. Princeton: Princeton University Press, 1992.

Mourelatos, A.P.D. "Aristotle's 'Powers' and Modern Empiricism." *Ratio* 9 (1967): 97–104.

Mueller, I. "Aristotle on Geometrical Objects." In *Archiv für Geschichte der Philosophie* 52 (1970): 156–71. Reprinted in *Articles on Aristotle*, vol. 3: *Metaphysics*, edited by J. Barnes, M. Schofield, and R. Sorabji, 96–107. London: Duckworth, 1979.

———. "Greek Mathematics and Greek Logic." In *Ancient Logic and Its Modern Interpretations*, edited by J. Corcoran, 35–70. Dordrecht, Neth.: D. Reidel, 1974.

———. *Philosophy of Mathematics and Deductive Structure in Euclid's "Elements."* Cambridge, Mass.: MIT Press, 1981.

———. "On the Notion of a Mathematical Starting Point in Plato, Aristotle, and

Euclid." In *Science and Philosophy in Classical Greece*, edited by A.C. Bowen, 59–77. New York: Garland, 1991.

Nagel, E. *The Structure of Science: Problems in the Logic of Scientific Explanation*. New York: Harcourt, Brace, and World, 1961. Reprint. Indianapolis: Hackett, 1979.

Owen, G.E.L. "'Tithenai ta Phainomena.'" In *Aristote et les problèmes de méthode*, edited by S. Mansion, 83–103. Louvain: Publications Universitaires de Louvain, 1961. Reprinted in *Articles on Aristotle*, vol. 1: *Science*, edited by J. Barnes, M. Schofield, and S. Sorabji, 113–26, London: Duckworth, 1975; and in G.E.L. Owen, *Logic, Science, and Dialectic*, edited by M.C. Nussbaum, 239–51. Ithaca, N.Y.: Cornell University Press, 1986.

———. "Aristotle on the Snares of Ontology." In *New Essays on Plato and Aristotle*, edited by R. Bambrough, 69–95. New York: Humanities Press, 1965. Reprinted in G.E.L. Owen, *Logic, Science, and Dialectic*, edited by M.C. Nussbaum, 259–78. Ithaca, N.Y.: Cornell University Press, 1986.

Pacius, J. *In Porphyrii Isagogen et Aristotelis Organum, Commentarius Analyticus*, Hildesheim, Ger.: Olms, 1966.

Patzig, G. "Erkenntnisgründe, Realgründe, und Erklarungen (zu *An. Post.* A 13)." In *Aristotle on Science: Proceedings of the Eighth Symposium Aristotelicum*, edited by E. Berti, 141–56. Padua: Editrice Antenore, 1981.

Pellegrin, P. "Aristotle: A Zoology without Species." Translated by A. Preus. In *Aristotle on Nature and Living Things: Philosophical and Historical Studies Presented to David M. Balme on His Seventieth Birthday*, edited by A. Gotthelf, 95–115. Pittsburgh: Mathesis Publications, 1985.

———. *Aristotle's Classification of Animals*. Translated by A. Preus. Berkeley: University of California Press, 1986.

Philoponus, J. *In Aristotelis Analytica Posteriora Commentaria*, edited by M. Wallies. No. 13, vol. 3, *Commentaria in Aristotelem Graeca*. Prussian Academy Edition. Berlin: George Reimer, 1909.

Putnam, H. *Mind, Language, and Reality: Philosophical Papers, Volume 2*. Cambridge: Cambridge University Press, 1975.

Ross, W.D. *Aristotle's "Prior and Posterior Analytics."* Oxford: Clarendon Press, 1949.

———. ed. *The Works of Aristotle Translated into English*. Oxford: Clarendon Press, 1928.

Russell, B. *Essays in Analysis*, edited by D. Lackey. London: Allen and Unwin, 1973.

Rutten, C. "Note sur le syllogisme de l'essence chez Aristote." *Phronesis* 9 (1964): 72–81.

Scholz, H. "The Ancient Axiomatic Theory." Translated by J. Barnes. In *Articles on Aristotle*, vol. 1: *Science*, edited by J. Barnes, M. Schofield, and R. Sorabji, 50–64. London: Duckworth, 1975.

Sellars, W. *Philosophical Perspectives: Metaphysics and Epistemology*. Springfield, Ill.: C.C. Thomas, 1967.

Smith, R. *Aristotle: "Prior Analytics."* Indianapolis: Hackett, 1989.

Smyth, H.W. *Greek Grammar*. Revised by G. Messing. Cambridge, Mass.: Harvard University Press, 1956.

Sorabji, R. "Definitions: Why Necessary and in What Way." In *Aristotle on Science: Proceedings of the Eighth Symposium Aristotelicum*, edited by E. Berti, 205–44. Padua: Editrice Antenore, 1981.

Themistius. *Analyticorum Posteriorum Paraphrasis*, edited by M. Wallies. No. 5, vol. 1, *Commentaria in Aristotelem Graeca*. Prussian Academy Edition. Berlin: George Reimer, 1900.

Tredennick, H., trans. *Aristotle: "Posterior Analytics" and "Topics."* Loeb Classical Library. Cambridge, Mass.: Harvard University Press, 1960.

Tricot, J. *L'Organon d'Aristote, traduction nouvelle et notes*, 2d ed. Paris: J. Vrin, 1979.

Upton, T. "Imperishable Being and the Role of Technical Hypotheses in Aristotelian Definition." *Nature and System* 2 (1980): 91–100.

———. "The If-It-Is Question in Aristotle." *Ancient Philosophy* 11 (1991): 315–30.

von Fritz, K. "Die APXAI in der griechischen Mathematik." *Archiv für Begriffsgeschichte* 1 (1955): 13–103.

Zabarella, J. *Opera Logica*. Hildesheim, Ger.: Olms, 1966.

Index of Names

Ackrill, J.R., 65n.48, 115, 115n.26, 117, 125n.44, 131n.58, 132n.62, 133n.63, 157
Aquinas, T., 47nn. 9, 11, 67n.52, 106nn. 9, 10, 111n.22, 117, 125n.46, 128n.52, 134n.65, 157

Bäck, A., 73n.65, 157
Balme, D., 3n.6, 50n.17, 157
Barnes, J., vii, viii, 1n.2, 2n.4, 4n.8, 17n.2, 18n.4, 20n.7, 21, 22n.11, 26, 33n.23, 47n.8, 48, 54n.26, 55, 77n.70, 80n.3, 83n.4, 84n.6, 86n.9, 87n.12, 88n.13, 89n.15, 95n.23, 109, 111n.21, 117, 123n.41, 134n.64, 142, 142n.6, 146, 151n.16, 157
Bolton, R., 66n.51, 98n.29, 105n.8, 114n.24, 122n.37, 124n.42, 128n.51, 132n.62, 157
Bonitz, H., 54n.26, 157
Broadie, S., viii, 65n.48, 66n.49, 75n.68, 93n.19, 118n.33
Burnyeat, M., 1nn. 1, 2, 15n.1, 32n.22, 157

Demoss, D., 85n.8, 98n.29, 133n.64, 158
De Rijk, L.M., 3n.7, 158
Deslauriers, M., 18n.4, 110n.19, 134n.64, 158
Devereux, D., 85n.8, 98n.29, 133n.64, 158

Euclid, 51n.20, 59, 60, 69, 97n.28

Ferejohn, M., 1n.1, 5–10, 14, 58n.34, 62n.41, 65n.47, 68n.54, 80n.2, 127n.48, 158
Frede, D., 71n.60, 158
Frede, M., 31n.21, 158
Freeland, C., 125n.47, 158
Frege, G., 69

Goldin, O., viii
Gomez-Lobo, A., 12, 17n.2, 41, 46, 52–58, 158
Gotthelf, A., 1n.2, 158
Graham, D., 139n.1, 158

Halper, E., 29n.19, 158
Hamelin, O., 107n.14, 158
Heath, T., 59nn. 35, 36, 150n.15, 158
Hempel, C., 153n.20, 158
Hesiod, 152n.17
Hilbert, D., 60n.37
Hintikka, J., 12, 41, 46, 50, 69–71, 97n.27, 158
Hussey, E., 73n.64, 74n.67, 158

Inwood, B., 144n.8, 146, 158

Jacobs, W., 63n.43, 64n.45, 158

Kahn, C., 21n.10, 53, 57, 63n.43, 159
Kim, J., 153n.18, 159
Klein, J., 51n.19, 159
Kripke, S., 147, 159
Kullmann, W., 5n.11, 159

Landor, B., vii, 43n.2, 53n.24, 101, 103nn. 4, 5, 105n.8, 111n.21, 123nn.

Landor, B. *(continued)* 39, 41, 124n.42, 128nn. 50, 51, 129n.54, 159
Lee, H.D.P., 59n.36, 159
Lennox, J., 1n.2, 7–8, 159
Lloyd, A.C., 124n.41, 159
Lukasiewicz, J., 62, 159

Mansion, S., 67n.54, 101n.1, 102, 104n.7, 106–8, 119n.34, 125n.46, 159
McKirahan, R., 3nn. 4, 6, 5–6, 9–10, 14, 17n.2, 18n.4, 22n.12, 26n.17, 33n.23, 43nn. 1, 2, 48nn. 12, 15, 50n.18, 54n.26, 59n.36, 60n.38, 61n.39, 77n.70, 101n.1, 110n.19, 115n.25, 140n.2, 142n.5, 146n.10, 159
Mendell, H., 62n.41
Mourelatos, A.P.D., viii, 55n.30, 97n.27, 118n.33, 130n.57, 159
Mueller, I., 2n.3, 50n.18, 60n.37, 61n.40, 73n.64, 159
Mure, G.R.G., 55

Nagel, E., 153nn. 19, 20, 160

Owen, G.E.L., 22n.12, 69n.55, 80n.1, 114n.23, 160

Pacius, J., 19n.5, 106n.10, 160
Patzig, G., 120n.35, 160
Pellegrin, P., 3n.6, 50n.17, 87n.12, 160
Philoponus, J., 13, 101, 102–7, 110n.20, 111n.22, 160
Plato, 51, 87–88, 113, 153
Putnam, H., 147, 160

Ross, W.D., vii, 13, 17n.2, 18n.4, 19n.5, 20n.7, 23, 26–29, 33n.23, 36, 43n.1, 47n.9, 51n.20, 54, 72n.62, 74nn. 66, 67, 77n.70, 80n.3, 86n.9, 89n.15, 91n.17, 92n.18, 101, 105n.8, 111nn. 21, 22, 116n.28, 117, 119n.34, 121n.36, 125n.46, 128n.51, 129n.54, 130n.56, 133n.64, 134, 143n.7, 150n.15, 160
Russell, B., 64n.46, 69, 160
Rutten, C., 108n.16,160

Scholz, H., 48n.13, 160
Sellars, W., 153, 160
Smith, R., 2n.3, 87n.11, 160
Smyth, H.W., 43n.3, 160
Sorabji, R., 98n.29, 124n.42, 132n.62, 133n.64, 160

Themistius, 13, 101, 134–36, 161
Tredennick, H., 43nn. 1, 2, 161
Tricot, J., 19n.5, 93n.18, 110n.20, 111n.22, 161

Upton, T., 20n.7, 43n.3, 53n.24, 114n.23, 142n.4, 161

von Fritz, K., 54n.26, 161

Xenocrates, 86

Zabarella, J., 115, 161

Index Locorum

Categories
3 1b16–24: 147
10 11b38–12a2: 64
12a4–9: 64n.44
13a37–b3: 63
13b12–35: 63
13b20–21: 64n.45
13b20–27: 64n.45
13b28: 64n.45
Eudemian Ethics
2.1 1220a15–22: 117n.31
On the Heavens
4.4 311b4–9: 141
On Interpretation
6 17a20–26: 64n.45
7 17a39–40: 145
10 19b5–12: 64n.45
11 21a25–28: 65n.48
14 23b21–25: 16
Metaphysics
3.3 998b22–27: 95n.23
4.1 1003a21–22: 95n.23
5.1 1013a14–17: 47n.7
1013a18–20: 47n.7, 54n.26
5.2 1013a24–26: 47n.10
1013b20–21: 47n.10
5.7 1017a7–8: 131n.61
1017a22: 131n.61
1017a31–35: 45n.6
5.12 1020a4–6: 133n.63
5.18 1022a14–18: 73n.65, 76
1022a14–22: 5, 145n.9
1022a24–36: 145n.9
1022a27–29: 145n.9
1022a32–35: 76, 135
5.24 1023a35–b2: 48
5.30 1025a14–27: 20n.8
1025a14–30: 116n.28
1025a30–32: 5
6.1 1025b30–1026a6: 149
7.2 1028b28–29: 34n.24
7.4 1029b27–28: 114
1030a7–11: 99n.30
1030b8–11: 99n.30
7.7 1032b1–2: 33, 34
1032b14: 37
1033a1–5: 149
7.8 1033b5–8: 32, 34
7.12 1037b28–1038a4: 3n.6
7.15 1039b23–27: 32
1039b27–30: 124n.41
7.16 1040a27–b2: 124n.41
1040b16–24: 95n.23
7.17 1041a6–10: 27
1041a10–11: 32, 34
1041a10–14: 28
1041a16–20: 30
1041a16–22: 29
1041a23–26: 28
1041a25–26: 32
1041a26: 27
1041a26–30: 106
1041a26–32: 27, 30
1041a28–32: 129n.53
1041a31–32: 31, 33, 34n.25
1041a32: 30n.20, 32, 33
1041b4–9: 31, 33
1041b5–9: 106
1041b9–11: 34, 107
1041b27–31
8.2 1043a2–28: 149
8.4 1044a27–29: 141n.3

1044b8–20: 33
8.6 1045a7–b7: 84n.6
1045a12–14: 99n.30
1045a33–b7: 3n.7
1045b1–7: 95n.23
9.8 1050a4–23: 37
9.10 1051b30–33: 154
11.3 1061a28–b3: 73n.64
12.1 1069a30–b2: 36
13.3 1077b17–20: 150
13.9 1086a15: 54n.26

Nicomachean Ethics
1.7 1098a16–18: 37n.27

Parts of Animals
1.1 641b23–30: 37

Physics
2.1 193a30–31: 37
193b7–18: 37
193b12–18: 35
193b13–18: 37
2.3 194b23–26: 47n.10
194b26–29: 139n.1
195a18–19: 47n.10, 54n.26
2.9 200a7–30: 141n.3
4.8 216b20–21: 67n.53

Poetics
20 1457a28–30: 99n.30

Politics
1290b25–37: 87n.12

Posterior Analytics
1.1 71a11–17: 41, 44
71a14: 45n.6
71a14–15: 51n.20
71a15: 74n.66, 131n.59
71a15–16: 45n.6, 50, 93n.20, 131n.59
71a16: 20n.7, 45n.6
71a17–24: 57
71a19–20: 57
71a24–27: 56
71a24–30: 113
71a27–28: 57
71a28–29: 56
71a30–33: 57
1.2 71b9–12: 21
71b9–16: 32
71b21: 111
71b29–33: 111
72a5–8: 2
72a14–16: 42
72a14–21: 79
72a14–24: 2, 24, 41, 42, 45, 70
72a16–18: 89n.15
72a18–24: 134
72a20: 54, 57, 67n.53
72a21–24: 74n.66
72a23–24: 84, 95
72a38: 46
1.3 72b18–20: 53
72b18–25: 79, 83
72b19–24: 2
72b23–25: 26
1.4 73a34–b16: 145n.9
73a35–38: 5n.13
73a36: 19n.5
73a38–b4: 5n.14
73b4–5: 5
73b5–8: 135
73b5–10: 22, 77n.70
73b7–8: 19n.5
73b10–16: 6–7, 20n.8, 135
73b13–16: 111
73b14–16: 7
73b16–18: 145
73b26–27: 145
73b28–29: 73n.85, 145
73b30–32: 51n.20
73b32–39: 144
73b32–74a3: 144
1.5 74a10–13: 144
74a36–74b4: 144
1.7 75a38–b20: 150
75a41–42: 42
75a42: 58
75a42–b2: 5
75a42–75b12: 42
75b1–2: 58
75b2: 74n.66
75b2–3: 42

75b4–5: 74n.66
75b5: 74n.66
75b8–9: 150n.15
75b11–12: 50n.18
75b14–17: 150n.15
75b17: 74n.66
75b18: 74n.66
75b21–36: 32
75b33–36: 33n.23, 125n.44
1.9 75b37–76a9: 150
76a9–15: 150n.15
1.10 76a31–34: 47
76a31–36: 24, 45, 70
76a31–77a4: 2
76a32: 93n.21
76a34: 70
76a34–35: 74n.66
76a35: 51n.20
76a35–36: 74n.66
76a36: 71
76a41–b2: 3n.4
76b1: 74n.66
76b2–5: 70
76b2–16: 70
76b3ff: 128n.51
76b3–6: 3
76b3–11: 49
76b3–16: 24, 41, 46
76b4: 70
76b4–5: 50, 74n.66
76b5: 50, 71
76b6: 49, 66n.50
76b6–7: 93n.21
76b6–11: 41, 70
76b7: 66n.50
76b7–8: 71
76b8: 51
76b9: 51, 71
76b11–13: 50
76b11–16: 50
76b12–15: 5
76b13: 41, 58, 71n.60
76b14–15: 3n.4
76b15: 66n.50
76b15–16: 41
76b16–18: 69
76b35–36: 54n.26
76b36: 54
76b39–77a3: 53n.24
1.11 77a26–32: 3n.4
77a26–35: 89n.15
1.12 77b5: 46
78a1–20: 17n.2
1.13 78a22–b4: 120
78a36–37: 127n.49
78a39–b3: 20
78a39–b4: 4
78a39–78b4: 127n.49
78b34–36: 150n.15
79a6–7: 19n.5
79a7: 19n.5
79a7–8: 60
79a7–10: 74
79a13–16: 151
1.14 79a24–29: 82
1.15 79a34–36: 2
1.22 82b37–83a1: 4
83a1–23: 22
83a1–83b31: 72n.62
83a24–b12: 19n.5
83b1–3: 4
83b17–20: 5
1.23 84b3–9: 51n.20
84b30–31: 43n.3
1.25 86b8–9: 43n.3
1.27 87a36: 74n.66
1.32 88b21: 46
88b28: 74n.66
88b29: 74n.66
1.33 89a19–21: 19n.5
89a32: 32, 71
89b7–8: 118
2.1 89b24: 54
89b24–25: 17
89b26–35: 17
89b32: 68
2.2 89b36–90a1: 18
89b37: 19
89b37–38: 35
89b38: 19

89b38–39: 54
89b39: 19
89b39–90a3: 22
89b39–90a4: 44
90a1–5: 55
90a2: 19, 84n.7
90a2–3: 18
90a3: 19
90a4: 19
90a4–5: 54
90a5: 26
90a5–6: 27
90a5–7: 119
90a5–11: 25
90a6–7: 35
90a7–15: 19
90a8: 19
90a9: 19
90a9–10: 18n.4
90a9–11: 22, 55, 108n.16
90a10: 54
90a12: 22
90a12–13: 54, 108n.16
90a12–14: 55
90a13: 19nn. 5, 6
90a14–15: 3, 21, 102, 109
90a14–23: 4
90a15–16: 129
90a15–17: 23
90a15–23: 21, 26
90a18–23: 4, 23
90a35–b1: 80
2.3 90b3–7: 82
90b7–17: 82
90b9–13: 83, 83n.4
90b16–17: 19n.5
90b20–24: 83, 83n.4
90b24–27: 46, 83
90b28–91a11: 83
90b33–38: 83
90b33–91a6: 85
90b35–38: 84
90b38: 84
90b38–91a2: 85
2.4 91a14–18: 86
91a18–25: 86
91a25–32: 86
91b7–9: 19n.5
2.5 91b12–17: 87
91b18–24: 88
91b18–26: 68n.54
91b23–24: 87n.12
91b24–26: 88
91b26–27: 88
91b28–35: 88
2.6 92a6: 19n.5, 89n.15
92a7–9: 88
92a9–10: 88
92a11–19: 89
92a12: 89n.14
92a20–24: 89
92a26–27: 90
92a27–33: 91
2.7 92a34–35: 19n.5
92b4–5: 93
92b4–8: 66
92b4–18: 93
92b5–8: 68n.54, 71
92b6–7: 67n.51
92b7: 68
92b8–11: 95
92b12–13: 93n.18
92b12–14: 19n.5
92b12–16: 95
92b12–18: 58n.34
92b14: 113
92b15–16: 51n.20
92b17: 93n.19, 95
92b17–18: 93, 95, 96
92b19–21: 96n.26
92b19–25: 97
92b21–22: 96n.26
92b23–25: 96n.26
92b26–28: 98
92b27–28: 66n.50
92b28–34: 98
92b29: 19n.5, 98n.29
2.8 93a1–3: 102
93a3–4: 102
93a3–9: 109

93a5–7: 105, 106
93a5–9: 102
93a6: 126
93a6–9: 82
93a9–13: 103, 111
93a9–15: 111n.22
93a14–15: 103
93a16–29: 106n.9, 112
93a16–b14: 104
93a21: 132n.62
93a21–29: 117n.31, 130
93a22–23: 124n.42
93a22–24: 117
93a23: 110n.19
93a23–24: 108
93a26–27: 113
93a27–28: 115
93a28–29: 113
93a29–37: 118, 120
93a35–37: 140
93a35–b7: 116
93a37–b7: 121
93a39–b7: 117
93b2–3: 23, 79
93b2–7: 4
93b5: 133n.64
93b6: 123n.39
93b7–14: 4, 124
93b8–12: 104
93b9: 110n.19
93b11: 110n.19
93b13–14: 125n.46
93b15–20: 126
93b18–19: 105
93b18–20
93b19: 16, 27, 105, 130, 138
93b19–20: 128n.50
93b20: 105n.8
2.9 93b21: 105n.8, 106
93b21: 27, 36, 105
93b21–28: 127
93b23–25: 130
93b25: 105n.8
93b25–26: 105n.8, 130
93b25–28: 19n.5, 138
2.10 93b29–32: 131n.58
93b29–35: 66n.50, 115n.25, 130
93b30–32: 66n.50
93b32: 131n.58
93b33: 131n.58
93b35–37: 132
93b38–39: 98n.29
93b38–94a7: 129n.53, 132
93b38–94a9: 4
94a2: 90
94a5: 129
94a7–9: 125n.45, 133
94a8–14: 66n.50
94a9–10: 134
94a11–14: 133n.64
94a12–13: 129n.55
94a12–14: 127n.49
2.12 96a10: 6
2.13 96a24–b14: 9n.23
96a34–35: 19n.5
96b6: 19n.5
96b28–97a6: 88
97a12–14: 19n.5
2.16 98b5–10: 24n.15, 139, 144
98b5–16: 4, 8
98b16–24: 4
98b25–31: 143
98b32–38: 143
98b38: 144
2.17 99a7: 143
99a8–11: 143
99a11–16: 143
99a12–15: 147n.13
99a23–25: 144
99a23–29: 4
99a24–25: 139
99a25–29: 8, 139, 141
99a28–29: 24n.15, 144
99a30–b3: 143
99a33–35: 144, 144n.8
99b5–6: 4
99b5–7: 4
2.19 99b20–22
100a5–6: 91

100b1–5: 91
100b5–15: 26

Prior Analytics
1.2 25a1–26: 62
1.23 40b25–29: 54n.26
1.23 41a23–27: 54n.26
1.30 46a26–27: 127n.49
46a33: 87n.11
1.35 48a29–39: 2n.3
1.44 50a16–b4: 54n.26
2.5 58a35: 127n.49
2.16 64b29: 127n.49

Sophistical Refutations
1.5 166b37–167a4: 18n.4

On the Soul
1.1 403a25–b16: 102
403a25–b19: 149
403b5–6: 30
2.1 412a22–27: 37

Topics
1.5 101b38: 133n.63
3.6 119b35–39: 54n.26
4.1 121a18–19: 95n.23
121b16: 95n.23
6.9 147a29–b25: 89
7.3 153a15–18: 3n.6
153a23–b24: 91n.17
153a26–b24: 89, 91n.17